GREENWICH OBSERVATORY

GREENWICH OBSERVATORY

One of three volumes by different authors
telling the story of Britain's oldest scientific institution

THE ROYAL OBSERVATORY
AT
GREENWICH AND HERSTMONCEUX
1675–1975

Volume 1: Origins and Early History (1675–1835)

by

Eric G. Forbes

UNIVERSITY OF EDINBURGH

ALERE FLAMMAM

TAYLOR & FRANCIS · LONDON

1975

First published 1975 by Taylor & Francis Ltd.,
10–14 Macklin Street, London WC2B 5NF.

ISBN 0 85066 093 9

Design and production in association with Book Production
Consultants, 125, Hills Road, Cambridge.
Set in 12/13 Monotype Bembo Series 270 by The Lancashire
Typesetting Co. Ltd., Bolton.
Printed by Taylor & Francis (Printers) Ltd., Rankine Road,
Basingstoke, Hampshire.
Bound and slip-cased by the University Printing House,
Shaftesbury Road, Cambridge.
Blocking brass design by William Andrewes.

To Andrea Christine

Preface

THIS is the first volume of a three-volume history written to commemorate the Tercentenary of the Greenwich Observatory. In it a description is given of the endeavours of its first six directors, or Astronomers Royal, whose periods of office span the 160 years from the time of its foundation until 1835. These were

1. John Flamsteed (1646–1719), Astronomer Royal from 1675 to 1719.
2. Edmond Halley (1656–1742), Astronomer Royal from 1720 to 1742.
3. James Bradley (1693–1762), Astronomer Royal from 1742 to 1762.
4. Nathaniel Bliss (1700–64), Astronomer Royal from 1762 to 1764.
5. Nevil Maskelyne (1732–1811), Astronomer Royal from 1765 to 1811.
6. John Pond (1767–1836), Astronomer Royal from 1811 to 1835.

The follow-up to this part of the narrative has been written by Professor A. J. Meadows of Leicester University, who ends the story with an outline of current astronomical researches at the Observatory's new site at Herstmonceux Castle in Sussex. The author of the third volume, Lieut.-Cdr. Derek Howse, Curator of Astronomy at the National Maritime Museum, has supplied detailed information about the various instruments employed at Greenwich, so many of which survive there in his charge, including the principles of their construction, the names of those by whom and for whom they were made, the manner in which they were used, etc. By showing where and when buildings were added to mount them or to store them, he has linked the growth of the Observatory's activities to its physical expansion throughout the three-hundred years of its existence.

In order to portray the intellectual scene which nurtured this development, it has been necessary in the present volume to focus upon only a few aspects of each Astronomer Royal's work. Although we have been relieved of the burden of providing the logistics of the major instruments mentioned, and discussing how they were used in conjunction with ancillary apparatus such as lenses, micrometers, clocks, etc., enough has been said about observational techniques to enable an intelligent lay reader to appreciate the significance of the results obtained. Another balance which we have sought to achieve is that of viewing the contributions of the Observatory both in an internal (personal) and external (international) light. Some distortions inevitably occur. I am conscious, for example, of having given comparatively little attention to the development of astronomical and navigational tables and to the computational methods used at different times to reduce the astronomical data collected. That would have been a time-consuming task, and perhaps rather too technical for this broad survey. Moreover, there is evidence that such matters were given low priority by the early Astronomers Royal themselves, partly, no doubt, because they lacked training in the powerful new analytical methods being introduced by French, German, and Swiss mathematicians and astronomers during the eighteenth century which revolutionized the scope and accuracy of lunar and planetary theory. Such a deficiency was recognized in 1832, towards the close of the period with which we are here concerned, by Sir George Biddell Airy who reported on it as follows to that year's British Association Meeting at Oxford:

"The characteristic difference between the spirit of the proceedings in England and on the Continent may be stated thus.—In England, an observer conceives that he has done everything when he has made an observation. He thinks that the merely noting the passage of a star over one wire and its bisection by another, is all that can be expected from him; and that the use of a Table of logarithms, or anything beyond the very first stage of reduction, ought to be left to others. In the foreign observatories, on the contrary, an observation is considered as a lump of ore, requiring for its production, when the proper machinery is provided, nothing more than the commonest labour, and without value till it has been smelted. In them, the exhibition of results and the comparison of results with theory, are considered as deserving much more of an astronomer's attention, and demanding greater exercise of his intellect, than the mere observation of a body on the wire of a telescope."

This situation was not allowed to persist in Airy's time since, unlike his immediate predecessor Pond, he possessed a sound knowledge of higher mathematics. It had, however, been prevalent throughout most of the first century-and-a-half of the Greenwich Observatory's history, and had also been criticized by Maskelyne in a memorial presented to the Council of the Royal Society in 1763. But nowhere is it more apparent than in the attitude of the first Astronomer Royal, John Flamsteed; indeed, it played a major role in the bitter conflict which he waged with eminent scientific contemporaries such as Newton and Halley, who were eager to reap the fruits of his lifetime of labour. He seldom allowed himself to become involved in any theoretical speculation until his large catalogue of reliable star positions should be complete, and died shortly before it was. He rightly regarded such data as a funda-mental necessity for improving the state of celestial navigation, which he had been commissioned by Charles II—the founder of the Observatory—to do. For his successor Halley, however, inspired by his knowledge of Newton's principle of universal gravitation, these were merely a means to an end of being able to predict the position of the Sun, Moon and planets. It was upon this complementary aspect of astronomy that Halley, and later Maskelyne, tended to concentrate most of their attention; although the latter also worked hard to establish the precise celestial coordinates of thirty-six bright stars useful in navigation. The impact of Bradley's numerous astronomical observations (and those of Bliss) was weakened by the tragic delay of roughly forty years in their publication.

It was Maskelyne's organizational ability which brought a much-needed stability to the Observatory and helped to regularize its relationships with the Board of Ordnance and Board of Visitors. The Board of Ordnance which, until 1763, had always paid the salaries of each Astronomer Royal and his assistant, was then surprised to find that it was also responsible for the upkeep of the buildings and the instrumentation. The Board of Visitors, consisting of the President and Council of the Royal Society and their nominees, had been established in 1710 to break Flamsteed's autonomous control of the Greenwich Observatory; but it was used by Maskelyne as an important line of communication to the Board of Longitude and hence to the Admiralty, through which the funds for the computation of the *Nautical Almanac* for 1767 onwards were to be provided. By Pond's era, the burdens of observational and computational work had grown to such an extent that it became necessary to greatly increase the staff. No longer was it practicable to administer the affairs of the Observatory through these three different channels. Thus, by the Act 58 George III, cap. xx of 1818, executive and financial responsibilities fell entirely into the Admiralty's hands. The Board of Visitors retained its useful role as the advisory body. Its constitution was radically revised in 1830 following William IV's granting of a Royal Charter to the Astronomical Society of London, after which the latter began to share this duty equally with the Royal Society; as indeed, it still does to this day.

This brief summary of how the Greenwich Observatory was transformed from a meagrely financed one-man institution struggling for survival under the negligent eye of the Merry Monarch, into a flourishing establishment with a director and six assistants paid from Admiralty funds and provided with expensive equipment, books, and other necessities, is expanded in the eight chapters which constitute the text. Chapter 1 is an attempt to indicate the background of scientific, and in particular astronomical, development which immediately preceded its foundation, and the serious problems facing oceanic navigation at that time. The latter aspect has been treated at considerable length, since many of the points mentioned are not often made explicit in conventional text-books. The immediate origins are shown to be an admixture of an urgent and long-standing need to solve a practical problem of national importance coupled with an element of charlatanism at the Court of Charles II. Chapter 2 is a study of the youthful intellectual development of Flamsteed, introducing the results of recent original research into his unpublished papers and correspondence. The treatment here given is meant to show why he interested himself in certain types of investigations at the beginning of his long term of office, and how he became a victim of circumstances. Chapter 3 explains the reasons for his clashes with Newton and others that led to the establishment of the Board of Visitors in 1710 and the pirating of his *Historia Coelestis* two years later; and how his own three-volume edition of that work eventually came to be published over five years after his death. These circumstances will be familiar to most specialists in the history of astronomy through Francis Baily's (1835) and William Cudworth's (1889) respective publications of Flamsteed's autobiographical notes and some of his later correspondence with his erstwhile assistant Abraham Sharp.

Chapter 4 describes the grounds of the long-standing quarrel between Flamsteed and Halley partly as an excuse for introducing aspects of their individual contributions to astronomy not otherwise represented in our story, but also because of what it reveals regarding their personalities. The origins of the Board of Longitude in 1714, and the contemporary concern with improving the predictive ability of the lunar theory, are also discussed. The significance of Halley's long series of observations of the Moon's position made at Greenwich towards the end of his life are finally assessed. The fact that other facets of Halley's versatile and interesting scientific career are readily accessible in several modern biographical studies justifies our omission of these from the text.

Chapter 5 begins by explaining the background and implications of Bradley's famous discoveries of stellar aberration and nutation, which were responsible for establishing him as a figure of international importance. Our subsequent account of the circumstances surrounding the long delay in publishing his and Bliss's Greenwich observations was based principally on a study of four volumes of Board of Longitude papers in the archives of Herstmonceux Castle. The manner in which Bessel utilized these data in the early decades of the nineteenth century for founding an exact science of positional astronomy is finally briefly described. The significance of what Bradley achieved during his twenty years at Greenwich is viewed in Chapter 6, alongside the results of Maskelyne's early scientific work, from an international perspective, and further examined in Chapter 7 in the context of a discussion of Maskelyne's improvements to Bradley's methods and instrumentation.

Chapter 6 describes the involvement of both these Astronomers Royal with the claim of the German astronomer Tobias Mayer for one of the Longitude prizes offered by the British parliament under the terms of the Act 12 Queen Anne 1714, and how the accuracy of the latter's solar and lunar tables inspired Maskelyne to found the *Nautical Almanac*; also with their respective contributions to the organization of the British expeditions to observe the transits of Venus in 1761 and 1769. Since no biography of Maskelyne has ever been written, we were obliged in preparing Chapter 7 to rely heavily upon his published articles in the *Philosophical Transactions of the Royal Society*. The lack of such readily accessible information regarding his contributions to the *Nautical Almanac*, and the existence of detailed Board of Visitors' manuscript minutes from 1763 onwards, helped to dictate our policy in Chapter 7 of biasing the discussion of his work at Greenwich more heavily towards his meridian transit observations than towards problems associated with the extra-mural management of those employed in computing this annual periodical. However, the problems which the latter had to face are not entirely neglected, and form a link with Pond's requirement for an increased staff. The same primary source materials have also proved to be of great assistance in framing our approach to the discussion of Pond's work at Greenwich to which Chapter 8 has been devoted.

This study has served to confirm how much still remains to be done before a definitive history of the Greenwich Observatory can be undertaken. There are thousands of documents in its archives which still require to be read and historically evaluated. This book makes no pretence to giving a comprehensive view of the subject, but should enable the reader to appreciate what was done during the first 160 years in the life of this highly respected institution to establish the astronomical tradition of which this country is now so proud.

14 *August* 1974   Eric G. Forbes

Acknowledgements

I T gives me great pleasure to acknowledge the assistance I have received during my preparation of this book from Lieut.-Cdr. Derek Howse and his assistant Mrs Valerie Finch of the National Maritime Museum. Without it, I might have overlooked some useful printed and xeroxed sources, and have failed to expedite my literary task as quickly as I have done. I was also able to profit from discussions with Mr Philip S. Laurie of the Royal Greenwich Observatory who for years has been unofficially responsible for the unique and priceless manuscripts in the archives of that institution, and is consequently very knowledgeable regarding their contents. The libraries to which I am most indebted for the secondary literature consulted are those of the University and Royal Observatory in Edinburgh and the Royal Society in London. My thanks are due to the Trustees of the National Maritime Museum, (Figures 8 and 12) the President and Council of the Royal Society, (Figures 9, 11 and 13) the Director and Trustees of the National Portrait Gallery, (Figure 10) for permission to reproduce the illustrations which complement the text. My work has been financed primarily by the Alexander von Humboldt Stiftung in Bonn, and supported by a research grant from the Faculty of Arts of Edinburgh University. The manuscript was conscientiously typed by Mrs Finch, and Miss Elizabeth Gibson of the History Department, Edinburgh University.

List of Illustrations

Contents

1

Introduction

BEFORE the practice of astronomy was revolutionized in the seventeenth century by the advent of the astronomical telescope and pendulum clock and the application of telescopes to quadrants and sextants, there existed few observatories worthy of the name. The Nuremberg home of Johannes Müller of Königsberg (better known as Regiomontanus), the castle of William IV, Landgrave of Hesse-Cassel, and the palace of Tycho Brahe on the small island of Hven near Copenhagen in the Danish Sound, contained diverse instruments for observing the altitudes of heavenly bodies, but the nominal accuracy of these measurements was far in excess of the value actually attainable; for example, $\pm 1''$ as compared with $\pm 5'$ in Tycho's case. The uncertainties accruing from the lack of a theory to compensate the effects of atmospheric refraction, an incorrect allowance of about $3'$ rather than $8\cdot 8''$ for the Sun's parallax, and the contemporary ignorance of other astronomical phenomena affecting the apparent positions of stars, were probably as much to blame for imposing these limits upon the precision of Tycho's data as errors in the design, construction and mounting of his instruments or in his observational techniques.

However, the chief value of Tycho's attempt to provide a reliable foundation for astronomical science lay in the continuity and extent of his planetary observations. It was the combination of these factors with the comparatively high accuracy of his data, and not merely the accuracy alone, which forced Johannes Kepler to acknowledge the reality of the discrepancy of $8'$ from perfect circularity in the shape of Mars's orbit and establish his first two laws of planetary motion; namely, that the orbits of planets are elliptical with the Sun at one of the two foci, and that the radius joining the planet to the Sun sweeps out equal areas of the orbit in equal intervals of time. Galileo probably felt that he could not accept

I

Kepler's inference that every planet revolved in an elliptical orbit without rejecting the principle of circular inertia upon which his entire system of mechanics was based, and so preferred to remain silent on this issue while supporting Kepler's adoption of the Copernican axioms of heliocentricity and diurnal rotation. Descartes's vortex theory of planetary motion was based on different axioms of motion demonstrated deductively from ontological arguments that were considered to carry much more weight than Kepler's empirically derived conclusions, and this was a further obstacle to the latter's acceptance. A popular alternative proposed by Ismael Bouilleau (alias Bullialdus) in 1645 was to give each orbit the metaphysical justification of being the actualization of an ideal mathematical form, and assign to the empty focus of the Keplerian ellipse the same significance as the equant point had been given in the Ptolemaic circular orbit theory, thereby offering what then appeared to be a simpler and aesthetically more satisfying model of planetary motion. More sophisticated equant theories capable of predicting a planet's position to within $\pm 1'$ were soon developed by Bouilleau himself in France and by Seth Ward in England, so nothing had to be sacrificed from the observational standpoint.

Isaac Newton was initially among those who preferred the equant theory to Kepler's second law. Since, however, he regarded the Earth's orbit around the Sun as being subject to modification by the Moon's vortex, he did not adhere to Bouilleau's precept that it had to be an exact ellipse. This is tantamount to saying that Newton initially refused to accept the validity of Kepler's first two laws of planetary motion; which is indeed a major reason why, after deriving the inverse-square distance dependence of gravitational attraction by applying Kepler's third law to the Moon's orbit in 1666, he did not invoke the hypothesis of universal gravitation as a causal explanation of the other two laws. Some twenty years were destined to elapse before he could overcome his philosophical belief in Descartes's vortices and his mathematical preference, as a geometer deeply steeped in the classical tradition of Euclid and Apollonius, for the equant theory of planetary motion. Only then was he able to demonstrate that the law of equal areas is a consequence of the single assumption that the total attractive force acts centrally (i.e. towards the Sun), and that if the orbit is elliptical, this centripetal force must vary as the inverse square of the distance. Conversely, it followed that a centrally directed inverse-square law of gravitational force constituted a sufficient rational basis of Kepler's first two—hence all three—laws of planetary motion. This is the famous result embodied in Newton's *Philosophiae naturalis Principia, mathematica,* or simply *Principia,* of 1687, which was to provide astronomers with a new starting-point for dynamical investigations of the solar system and dominate scientific and philosophical enquiry for more than two centuries to follow.[1]

While these theoretical strides were being painfully taken, new vistas were being opened up by the application of the telescope to the study of celestial phenomena. Galileo's pioneering observations reported in his *Sidereus nuncius* of 1610 quickly became known. Here he describes the general appearance of the Moon, the nature of the Milky Way, and the motions of Jupiter's satellites; soon afterwards he announced his discoveries of the phases of Venus, the appendages (or unresolved rings) of Saturn, and sunspots. Simon Marius, in Germany, independently discovered the satellites of Jupiter, and succeeded in deriving better values than Galileo for their periods of revolution and other elements of their orbits; he is also credited with having been the first to observe the Andromeda Nebula in 1612. In England the earliest known telescopic observer was Sir Walter Raleigh's tutor and friend, Thomas Harriot, who would appear even to have anticipated Galileo in the construction and use of astronomical telescopes. Owing to his failure to publish his discoveries, however, they were known only to a group of his friends. As soon as the celestial appearances revealed by the telescope became known, the knowledge which they provided was immediately assimilated by Marke Ridley with the earlier discoveries published in William Gilbert's *De Magnete* (1600)—an important work that had greatly influenced Kepler's own thought. Ridley's extension of Gilbert's magnetic explanation of the Earth's mobility brought him into bitter controversy with the Archdeacon of Salisbury (William Barlowe) who insisted, as the Lutheran editor of Copernicus's *De revolutionibus orbium coelestium* (1543) had done almost a century beforehand, that this motion ought to be regarded as a hypothesis for simplifying the astronomical calculations. However, a pro-Copernican tendency is reflected by the popular astrological almanacs which were bought and read by even the lowest classes among the literate population. The earliest of the English compilers of these annual almanacs who professed to being ardent supporters of the Copernican theory were Edward Gresham and Thomas Bretnor. The latter was very knowledgeable, and his writings had authority among the learned as well as among the multitude. Religious writers recognized the need to reconcile the new data with the essential tenets of the Christian religion rather than blindly oppose them; moreover, after Copernicus's work had been placed in 1616 on the Index of forbidden books, discussions on matters of this nature were being held at Gresham College in London by the professors of astronomy and geometry who generally supported the Copernican theory. Two years after Galileo's trial and official condemnation as a heretic by the Catholic Church, Henry Gellibrand proposed the Earth's motion as a possible cause of the secular change in the variation of the magnetic needle that his experiments had just revealed.

Meanwhile, in France, three distinct schools of amateur astronomical activity had arisen which may be broadly classified as provincial, parisian, and Jesuit. The first school, or group, included Joseph Gaultier, Guillaume du Vair, and Nicolas Claude Fabri de Peiresc, at Aix in Provence, who were concerned primarily with the application of the telescope to discovering hitherto unknown celestial objects or appearances. Peiresc, for example, discovered the Orion Nebula on 25 November 1610 and made regular observations of the planets. He determined the periods and motions of Jupiter's satellites and, with Gaultier's assistance, prepared tables of these motions in the hope that they might ultimately serve as a basis for longitude determination. To Gaultier belongs the credit of having sparked off Father Pierre Gassendi's enthusiasm for astronomy. Together with Peiresc, the latter installed a gnomon at Marseilles, organized a network of eclipse observations, and made a telescopic study of the Moon's major topographical features that were used by Claude Mellan for engraving a lunar map. Other provincial French gentlemen with whom the two were in correspondence over astronomical matters included Godefroy Wendelin, Jacques Valois, and Elzéar Féronce. Peiresc's death in 1637, followed by Gassendi's departure later that same year for Paris, disrupted this group's activities although there is some evidence to suggest that Honore Gaultier was following in his uncle's footsteps as late as 1652.

The Parisian school comprised mainly mathematicians and mechanicists, and the interest was accordingly more with the theoretical discussion of optical problems—in which connection

Descartes's *La Dioptrique* (1637) was a great inspiration to them—
than with actual observation. This group included Giles Persone de
Roberval, Etienne Pascal, Pierre de Carcavy, Bernard Frénicle de
Bessy, and Claude Mydorge. Meetings are known to have taken
place at the homes of several other gentlemen, notably Father
Marin Mersenne. Mersenne had a wide circle of correspondents
including Descartes, Gassendi, and Blaise Pascal, and did a great
deal to stimulate learned intercourse on an extensive range of subjects
in the new experimental philosophy.

The Jesuits were committed primarily to instruction, and a
large number of astronomical (and related) texts were published
under their auspices all over the world—even as far away as China—
Father George Fournier's *Hydrographie* (Paris, 1643) and Father
Philippe Labbe's *La Géographie royalle* (Paris, 1646) being but the
first two of many. Among those who were active in Paris around
the middle of the seventeenth century were Fathers Bertet,
Deriennes, Léotaud, Petau, and Regis; and in the provinces Pierre
Anthelme, Gilles Macé, Gabriel Mouton, Jean Tarde, and Pierre
Vernier.

It is commonly stated that the informal gatherings of mathematicians and philosophers in Paris at this time constituted the
intellectual origin of the French *Académie des Sciences* in 1666, the
initial intention being to form a stable and permanent group of
people who were knowledgeable not only in matters of science but
in history, literature, and the fine arts as well. This intention was
not realized; instead, only two sections emerged whose members
received pensions from King Louis XIV as well as financial
assistance with their researches: viz. *Mathematics*, which included
mechanics and astronomy, and counted among its founder members
Christian Huygens, Adrian Auzout, Carcavi, Frénicle, Jean Picard,
and Roberval; and *Physics*, which then comprised anatomy, botany,
chemistry, and physiology, and included Claude Bourdelin, Samuel
Cotterau Duclos, Jean Pecquet, and Claude Perrault. The first
secretary was Jean Baptiste Duhamel, and five younger men were
chosen as assistants. Huygens was elected President, and in his
opening address to the Physical Assembly he proclaimed his support
for the scheme outlined by Francis Bacon in his *Novum Organum*
(1620) characterized by an experimentally based inductive approach
and a utilitarian function.

In astronomy, the work of Auzout and Picard represented a

distinct advance, for it was they who formally introduced the practice of using telescopes in conjunction with graduated circles for the precise measurement of angles, and began making systematic use of micrometers for measuring the small angular separations of objects visible simultaneously in the field of view of the telescope. Picard also applied Huygens's recently developed pendulum clock to time the meridian transits of stars and thereby determine their differences in right ascension, in accordance with the practice first adopted at Cassel by the Landgrave William IV during the previous century. A special study was also made of the effect of astronomical refraction, the great importance of which for precision astronomy was just beginning to be appreciated.

Picard had long been aware that the time was now ripe for constructing planetary tables which would replace Kepler's *Tabulae Rudolphinae* (1627), with the aid of telescopes larger in aperture and as much as ten times longer than those (of 6 or 7 feet) used a few decades earlier in conjunction with an accurate pendulum clock. There was, however, the problem that the garden behind the Academy's accustomed meeting-place, where his astronomical observations were being carried out was hemmed in by houses; thus an appeal was made to the King to implement his earlier intention of founding a proper observatory. This was duly begun in 1667, according to the design of Perrault, and was effectively completed five years later.

Having outlined his desiderata for the observatory and established a working relationship with the Italian astronomer Giovanni Domenico Cassini who had been invited to Paris as director of that institution, Picard went in 1671 to Hven in order to determine accurately the position of Tycho Brahe's former observatory (Uraniborg) and thereby facilitate the intended comparison of the observational data used by Kepler with that about to be collected at Paris. The necessity of this visit was borne out by its result which revealed that Tycho had incurred an error of 18' in assigning the direction of the meridian. Another unexpected discovery was that the pole star appeared to exhibit a periodic motion, an observational fact which was not fully understood until over half a century later.[2] But in the long term, an even greater stimulus to the work of the Academy arising from Picard's trip was his introduction of the young Danish astronomer Ole Roemer who had assisted him in his Uraniborg investigations and who was to stun the members a few years afterwards by his unexpected discovery, from observations of the eclipses of Jupiter's satellites, that light travelled with a finite speed.

Another early expedition which was to supply an important and unexpected result was Jean Richer's journey to Cayenne where the principal object had been the accurate trigonometrical determination of the distance between Mars and the Sun. Although, strictly speaking, it proved to be a failure—Cassini's subsequent determination of this quantity as 25″ was based upon an alternative method involving that planet's diurnal parallax—Richer made the unexpected discovery that the length of a pendulum adjusted to beat seconds at Cayenne (latitude 4° 55′ N.) was shorter than at Paris (latitude 48° 52′ N). This was not due to differences in atmospheric temperature, although that is what is stated in the Paris Academy's *procès-verbal* of 27 April 1690, but evidence of the variable influence of gravity at different latitudes and of the Earth's being flattened at the poles—conforming to a prediction in Newton's *Principia* deduced from the Copernican assumption of our planet's 24-hour axial rotation, which was taken for granted. The return of Roemer to Denmark (1681), followed by the death of Picard one year later and the departure of Huygens after the revocation of the Edict of Nantes (1685), were great blows to astronomy within the Paris Academy. The consequent neglect in observational and laboratory work was accentuated by the lack of sympathy towards purely theoretical researches shown by Colbert's successor Francois Louvois after his appointment as Protector of the Paris Academy (1683), and the activities of the members in general waned until it was radically re-organized and enlarged in 1699.

In contrast, the morale of the members of the Royal Society of London during the closing decades of the seventeenth century would appear to have remained high. The precise origins of this society have been a matter of debate,[3] but its closest antecedent, both as regards its aims and purpose, was the Experimental Science group formed at Oxford during the Civil War, which included the eminent mathematician and divine John Wallis, John Wilkins (later Bishop of Chester), and the physician Jonathan Goddard who had previously attended weekly meetings at Gresham College in London where questions of natural philosophy had been discussed. The Oxford society met for a time at the lodgings of Robert Boyle before many of its key members returned to London following the Restoration of Charles II and resumed weekly meetings at Gresham

College. It was after one of these that a plan for establishing a society devoted to the pursuit of experimental knowledge along the lines prescribed by Francis Bacon in his *New Atlantis* (1627) was worked out, and duly realized with the foundation of the Royal Society by Charter on 15 July 1662. Its privileges were extended by a second charter granted one year later.

The Royal Society's early concern with problems of importance to navigation was evidenced during the very year of its inauguration by its setting up of a three-man committee to investigate the practicability of finding position at sea from measurements of the secular change in the variation of the magnetic needle. This committee consisted of Sir Robert Moray as Chairman, Lord Brouncker as President of the Royal Society, and Henry Bond as the man responsible for having drawn attention to this possibility in his editions of John Tapp's *Seaman's Kalendar* between 1636 and 1657. The mathematical practitioners William Marr, John Pell, Henry Phillippes, and John Seller were asked to check Bond's predictions by making an annual examination of a sundial in Whitehall Gardens, London. But after these experiments had been interrupted by the Great Plague and Fire of London, Bond submitted his table of predictions up to the year 1716 for publication in Henry Oldenburg's *Philosophical Transactions* for 1668.

By this time everyone who had a knowledge of maritime affairs was well aware that all the other known means of determining position at sea when out of sight of land for any length of time were far too unreliable for the purpose of accurate navigation. The very act of steering a heavy sailing-ship with a massive tiller or a whipstaff to control the rudder, particularly when she was being buffeted by strong winds on the high seas, was an arduous task for the most experienced helmsman; so much so, that it required all his skill and strength to hold a course to the nearest half-point (or between 5° and 6°) on the ship's compass, and there was seldom any need to have the latter graduated to within the standard thirty-two points. To make matters worse, the design and construction of steering compasses were notoriously poor, and the practice of compensating for the irregular variation from the true North in the direction of the magnetic needle at the place of manufacture was highly unsatisfactory. So-called azimuthal compasses, with which magnetic bearings could be taken, enabled this magnetic variation to be estimated to around 2°, but appear to have been little used except

8

upon East India Company ships. Experience had revealed that such measures were of little practical value for indicating longitude differences in the North Atlantic. As Edmond Halley was to show (cf. Chapter 4), the underlying reason for this is that the lines joining points of equal variation on a magnetic chart—the isogonics —tended to run in an east–west direction.

The distance sailed was normally estimated by seventeenth-century British navigators from timed measurements of the ship's speed made with a sand-glass and knotted log-line attached to a flat wooden quadrant (or log). The log was weighted in such a way that when hove into the sea it floated upright facing the ship, thereby resisting any tendency to be pulled towards the ship as the line was payed out. Knots were secured to the line at every seven fathoms (that is every 42 feet) so that the number passing through a mariner's hands at the end of half-a-minute could be determined even in the dark, and was taken to be equal to the ship's speed expressed in knots (or nautical miles per hour). This spacing of the knots corresponded to a traditionally accepted value of 5000 feet for a nautical mile to which sailors still adhered long after Jean Picard's measurement of the meridian arc between Mahoisine and Amiens in 1671 had established 6080 feet as the correct measure. By persistently adopting a value for the nautical mile which was about 18 per cent too small, the navigator was systematically over-estimating the distance sailed.

This error was, however, compensated by cumulative errors involved in the traditional practice of plane (or plain) sailing, in which the Earth's surface was treated as being flat. This was a good practical reason why the more precise practice of Mercator navigation, in which the Mercator chart was used, was not followed by most British navigators during the sixteenth and seventeenth centuries despite the fact that Edward Wright in his popular book *Certaine Errors in Navigation* had done much to make it practicable. The practising navigator was moreover well aware that the erroneous representation of coastlines on Mercator charts often nullified the theoretical advantages of using them in coastal waters. The direction of the prevailing (Trades or Westerly) winds, along a great circle arc on the terrestrial globe, was the controlling factor in a safe and swift Atlantic voyage; it being merely coincidental that the trade routes between Europe and the North American mainland colonies, the Caribbean Islands, and the East Indies, on which at this time a rapidly expanding proportion of Britain's trade was centred, approximated to great circles and hence to the shortest passages on the terrestrial globe.

By heaving the log hourly or two-hourly, the navigator could estimate the average speed of his ship through the water, disregarding changes of wind and weather in the interval. The direction in which the log drifted gave an indication of the ship's leeway but errors caused by waves striking the log, the log being dragged by the line, and the effects of temperature and humidity changes on the flow of sand through the glass, could not be assessed any more than could the contribution of currents in carrying the ship along without altering its velocity relative to the sea itself. The authors of contemporary text-books therefore made no attempt to analyse such uncertainties, and simply attributed the resultant error in estimating position by this means to ocean currents alone—a procedure, or convention, which tended to exaggerate their effects.

One way in which the oceanic navigator could test the accuracy of his estimates of his ship's position from the course steered and distance sailed—a procedure known popularly as 'dead-reckoning'— was by keeping a regular check upon his latitude. Ideally, his calculated latitude should have agreed with that derived from observations of the Sun's meridian altitudes, any difference being attributable to the net effect of the above-mentioned uncertainties inherent in his art. In the case of great-circle sailing, these could be compensated from time to time at a number of intermediate stages by making slight alterations to the course. More commonly, however, a ship would be steered so as to reach a point in the same latitude as, but well to the seaward of, the intended landfall; whence the navigator 'ran down' that latitude until his haven was sighted. This practice was preferred because it was safe, and simple to apply.

A further difficulty with which the mariner had to contend in the application of this method was the unreliability of the conventional Davis quadrant or backstaff for measuring the altitudes of the Sun or stars. In the first place, this wooden instrument could have been badly constructed or even warped. The failure of seventeenth-century astronomers to use Edward Wright's tabulated corrections for the effects of the atmospheric refraction and the dip of the horizon, caused them to over-estimate the Sun's observed meridian altitudes and hence (since the latitudes of their respective land-based observatories were always known) the solar declinations tabulated in seamen's almanacs. A similar failure on the part of contemporary mariners faced with the converse problem of employing these declination tables to calculate the latitudes at regular intervals during a voyage, resulted in systematic under-estimates being made in this coordinate. Moreover, they generally failed to take account of the change in the solar declination during the time interval corresponding to the difference in the ship's longitude east or west of the meridian for which these tables had been prepared. For such reasons, therefore, a single latitude determination could easily have erred by nearly $\frac{1}{4}°$, enough for a navigator to miss sighting small low islands such as the Barbados even though he considered his ship to be sailing along their known parallel of latitude!

Yet the greatest source of error in the seventeenth-century art of navigation was that of determining with accuracy the distance sailed in an east–west direction. There existed no precise means of measuring a ship's longitude difference with respect to the meridian of a land-based observatory or the port of departure; but if both this coordinate and the corresponding value of the latitude were known, the position at sea would be uniquely specified. Since the Earth rotates on its axis once in every twenty-four hours, this problem was solvable by a comparison between the mariner's local time and that at the same instant on some standard meridian; the result being quickly convertible into units of angular measure simply by making use of the identity 24 hours $= 360°$.

In principle, after balance-springs were introduced into pocket-watches during the 1670s, a navigator could have timed the Sun's transit of the meridian and applied the tabulated declination for the date in question to the altitude observed with a Davis quadrant, and thus obtain the value of the co-altitude. The Sun's co-altitude (or zenith distance) and co-declination also being known, he could then have solved a spherical polar triangle for the local hour angle and hence local apparent time of his observations. In practice, however, he more often found the instant of meridian transit by interpolating the times on his pocket-watch before and after noon, when the Sun appeared to be at the same height above the horizon. Theoretically, the result ought to have been 12^h oo^m oo^s, but it usually differed on account of the watch-error. A repetition of these equal altitude observations on successive days enabled the rate at which the watch was gaining or losing to be quantitatively determined. An error of less than ± 4 minutes per day was regarded as tolerable, since proportional differences could be added to, or subtracted from, the measured times of other celestial events.

More popular still was the time-honoured method of finding the ship's latitude and local time from observations of the altitude of the Pole star. This star appears to move daily around a tiny circle centred on the north celestial pole; thus its observed altitude never differs from the latitude by more than a small angle. The correction to be applied to the former in order to obtain the latter was discussed by William Bourne, Edward Wright, and other sixteenth-century writers on navigation. John Seller, a century or so later, continued their practice of giving a 'Rule of the North Star', yet in popular eighteenth and nineteenth-century navigation manuals by John Robertson, Andrew Mackay and John Norie, instructions for making such a reduction are no longer to be found. This silence may indicate that the rule was so well known that it was unnecessary to publish an explanation of it. Alternatively it may reflect the growing popularity of the double-altitude method of which the equal altitude method referred to above may be regarded as a special case. Whatever the reason, it would seem that it was not so much the method as the instrument for making the necessary altitude measurements that had to be improved before latitude at sea and local time could be reliably determined.

But what was to be done about the complementary and much more troublesome problem of finding longitude at sea and standard time? The idea of using a mechanical timekeeper for this purpose was apparently first proposed by Rainer Gemma-Frisius in his *De principiis astronomiae et cosmographiae* (Antverpiae, 1530) although the first instrument specifically designed for use at sea was Christian Huygens's pendulum clock, completed in the year 1660 with the help of a Scottish political exile then living in Holland, Alexander Bruce, the second Earl of Kincardine. After Charles II came to the throne, Bruce returned to Britain taking two such clocks with him which were tested on board one of the king's yachts in the presence of Robert Hooke, F.R.S. Two years later, they were taken on a naval mission to Guinea, but reports of their performances during this voyage were conflictory. Huygens then made arrangements for other models, which were weight-driven and regulated by pendulums swinging in a cycloidal path, to be tried on ships of both the British and French navies; and these were so highly regarded that two Fellows of the Royal Society decided to revise, enlarge and publish their inventor's instructions as to how they were to be applied to the longitude problem. However, it soon became apparent that pendulum clocks could never operate satisfactorily in a sailing-ship. Another difficulty that had been inferred by Jean Richer in 1672 from the above-mentioned experiments made at Cayenne, was that their periods of oscillation vary with latitude as a result of the changes in terrestrial gravity associated with the Earth's departure from sphericity. This opinion was duly confirmed ten years later by a French expedition sponsored by the Paris Academy of Sciences, which sailed from a small island off West Africa to Guadeloupe and Martinique in the West Indies. Although the hope for the future seemed to lie in Robert Hooke's controversial invention of the balance-spring as a substitute for the pendulum, no further progress was to be made in this direction until well into the eighteenth century, a theme to which we will consequently not be returning until Chapter 6.

One could, however, dispense with a mechanical timekeeper for carrying standard time to sea by substituting an astronomical 'clock' in its stead. The idea of using the eclipses of Jupiter's satellites as the basis for a method of longitude determination had occurred to Galileo in 1610, not long after their discovery. Jupiter is sufficiently remote from the Earth for the effect of parallax to be neglected, and the eclipses recur at frequent intervals, particularly in the case of the first or largest satellite. The facts that Jupiter is masked from our view for over six weeks of the year by the rays of light from the Sun, and that the observations can only be made telescopically, are two major disadvantages opposed to the adoption of this method for use at sea. The complexity of the theory of the satellites' motions later came to be recognized as a serious obstacle to achieving the requisite degree of accuracy in predictions. On the other hand, this had little influence on the use of this phenomenon as a means of finding the longitudes of astronomical observatories, sea-ports or headlands, since two independent observers situated on widely separated meridians could each record the instants at which the same eclipse occurred and subsequently compare notes to establish their difference in longitude. A difficulty which could not be satisfactorily overcome was that Jupiter's dense atmosphere causes the satellites to fade gradually from sight in a manner dependent upon the magnification and resolving power of the telescope, thereby giving rise to appreciable uncertainties in the estimated time of eclipse. Solar and lunar eclipses, and transits of Mercury and Venus, while providing a useful means of improving our knowledge of the Sun–Earth distance, were too infrequent to be of any value in the determination of longitude at sea.

The swiftest-moving celestial body which alters its position with respect to the background of stars is, of course, the Moon. Thus, if the mariner were to be supplied prior to his voyage with a reliable prognostication of when our satellite would be seen at a land-based observatory to eclipse the Sun, occult or approach a star, or cross the meridian, he could observe the same phenomenon at sea and calculate the local time at which it occurred in the manner indicated above. The difference between that and the calculated observatory time was then a direct measure of his longitude separation from that fixed meridian. Lunar eclipses and occultations are unfortunately too infrequent for the *mariner's* purpose, although they provide valuable checks on the Moon's position which assist the *astronomer* in perfecting the lunar theory. Prior to the publication of Newton's *Principia* in 1687, however, and indeed for some eighty years afterwards, the state of man's knowledge regarding the cause and precise amount of the numerous observed inequalities in the Moon's orbital motion was quite inadequate to render either the predictions of the standard time of the Moon's transit across the meridian or the so-called method of lunar distances (or lunars) practicable.

The principles of this last-mentioned method had already been suggested early in the sixteenth century by Amerigo Vespucci, Johannes Werner, Peter Apian, and Gemma-Frisius, and revived over a hundred years later by Kepler, Christian Longomontanus, and Jean Baptiste Morin. Morin, Regius Professor of Mathematics and Medicine at the University of Paris, submitted a more general version of Kepler's method to Cardinal Richelieu who ordered on 6 February 1634 that it be examined by a scientific committee which included Pascal, Mydorge, Boulenger and Beaugrand. Less than two months later, Morin had defended his proposal before these commissioners who were convinced of its potential utility but impressed neither by its originality nor its practicability. The measurements which the mariner required to make with his quadrant were the Moon's distance from the Sun or a bright zodiacal star and the altitudes both of that celestial body and the Moon, all taken (ideally) at the same instant and repeated several times to reduce the influence of accidental errors. Another source of uncertainty which Morin's predecessors had neglected was the reduction of the altitude observations to compensate the effects of atmospheric refraction and diurnal parallax.

By 1673, however, Henry Bond had computed elaborate tables of both the variation and the dip of a magnetic needle from which a network of what were later to be christened isogonal and isoclinal lines could be drawn over our globe, thus providing the basis of an alternative terrestrial method which appeared to rival any of these celestial ones. Thus Charles II appointed another committee to investigate Bond's claim to have discovered a law by which the Earth's magnetic poles required to move in order to produce the measured changes in magnetic variation. In a manuscript written by its secretary, Dr John Pell, in the British Museum,[4] one finds a 26-point précis of the historical line of development up to 25 May 1673 accompanied by the following assessment of the contemporary situation:

> "The present state of our knowledge of the magnet seems to be this. By Mr. Bonds Hypothesis and Tables of equal motions of the magnetic pole we may find the place of the magnetic pole in the triangle LMN where L signifies London, M the magnetic Pole and N the astronomical North Pole of
> the same earths equator. O some other place given by its longitude and latitude from London. To calculate what respect a needle at O has to the point M; that is what variation it has and what inclination under its own horizon. And contra having the variation and inclination at O the position of O in respect of N & S of the meridian NL.
> Bond saith by his table of inclinations and on inclination after made in any

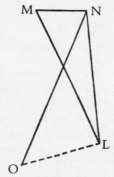

> place of the earth's face and the place's latitude, he will find the longitude of the place. But he forgets that he must also know the time (or at least the year) say not the hour of the observation. For without doubt in another year, in the latitude of London, some other point far from London may have the same inclination that now London has.
> I do not expect that an observation at sea can give me an inclination of the needle to 1/60 of a grade, but if it can be done on land it will

serve to place all islands and prominences, and landmarks on the coast in their true places in a chart. Having the longitude and latitude of Lands End and St. Helena, I can calculate the length of an arch of a great circle between Lands End and St. Helena. I can calculate all the angles which the arch makes with the intermediall meridians. But cannot by any means sail according to those angles (and thereby keep the *via brevissima inter duo puncta* proposition). But if Bond teach me how to calculate all the intermediate variations in every point of that arch, I can turn my astronomical angles into magnetical angles and so prescribe them to my steersmen.

But I have reason to believe that no steersman can follow my prescription and therefore, I must have my *new Fashioned Compass* where I can place my needle as much awry every day as I believe the compass varies, and so the finger above shall always be kept to the line or marked point in the edge of the box as if the steersman always steered precisely north. But if he desire to name winds or other mark, he must have another old-fashioned compass with all its winds."

This memorandum would appear to indicate that Pell was sceptical of the practicability of Bond's proposal for use at sea.

Less than three months later, Letters Patent bearing the date 19 August 1673 were issued by Charles II to the High Treasurer, Chancellor, chamberlains and barons of the Exchequer, to the officers of the Revenue and to the Mayor and Commonalty as Trustees of the property of the Royal Hospitals. These letters ordained that Christ's Hospital should be provided with a schoolmaster responsible for training forty poor boys

"whoe having attained to competence in the Grammar and Comon Arithmetique to the Rule of Three [viz. direct proportion] in other schooles of the said Hospitall may bee fitt to bee further educated in a Mathematicall Schoole and there taught and instructed in the Art of Navigaсon and the whole Science of Arithmetique until their age and competent proficiency in these parts of the Mathematiques shall have fitted and qualified them in the judgment of the Master of the Trinity House for the tyme being to bee found out as Apprentices for seaven yeares to some Captaines or Comanders of Shipps."[5]

The Governors of Christ's Hospital were to provide the necessary books, globes, maps, and mathematical instruments for the boys' instruction, and even clothes on their being accepted as apprentices.

Neither the conception itself nor the £7000 endowment for this project came from Charles II. A wealthy benefactor of that hospital, Sir Robert Clayton, initiated the idea and elicited the influential support of the Lord Treasurer (Clifford), Surveyor-General of the Ordnance (Sir Jonas Moore), and the Secretary of the Admiralty (Samuel Pepys). Moore subsequently induced the Lord High Admiral (the youthful James, Duke of York) to mediate for them. The money represented part of a legacy willed to the hospital by the former Governor Richard Aldworth almost thirty-seven years previously, which had been invested for most of that time in precarious Government Securities. The war with Holland which had been waging since Charles had come to the throne had incurred a serious loss to the complement of naval officers, so it was natural that the King and his naval authorities should be eager at that time to establish a regular means of thus supplying the nation's needs. However, only months after the Royal Mathematical School at Christ's Hospital had been established, the Dutch War came to an end and the King's (or 'Blue-Coat') Boys were alternatively employed in the mercantile marine.

The Master was expected to be sober, discreet, diligent, clean-living, well-organized, and a good expositor. Academically, he was to be skilled both in writing and conversing in Latin and Greek, encourage the boys by example to write neatly, and be a skilled mathematician able to justify himself in the face of questioning by foreign scholars or local practitioners. The first person elected, John Leake, was deemed unsatisfactory for quite a different reason: he tried to augment his rather meagre salary by taking private pupils, and neglected the Blue-Coat boys. Sir Jonas Moore was proving the sincerity of his interest in this training scheme by beginning a series of class-books on Practical Geometry, Trigonometry, and Cosmography; and Leake's replacement, Peter Perkins, was able to assist him with his Algebra, his Euclidean Geometry, and Navigation.[6] But Sir Jonas's concern with the unsatisfactory state of this last-mentioned subject went beyond the bounds of his literary endeavour on behalf of the Royal Mathematical School. He was also deeply interested in astronomical science, not least on account of the great relevance of such knowledge to navigational theory, and was a member of the above-mentioned committee set up by Charles II in 1673 to re-investigate Bond's theory of magnetic variation, the others being Lord Brouncker, Robert Hooke, Sir Robert Moray, Samuel Morland, John Pell, Sir Charles Scarborough, Colonel Silius Titus, Seth Ward, and Sir Christopher Wren.

While that committee was still considering this important matter, a Frenchman at Charles's Court and protégé of Louise de Kéroualle, Duchess of Portsmouth—the Sieur de St Pierre—announced that he had a lunar method for finding the longitude. The king therefore ordered his Secretary of State to issue the following Warrant to its members:

"Com^t to Ld Brouncker
&c about finding out the Longitude

Charles the Second &c To Our R^t Trusty & Wellbeloved Wm Ld Viscount Brouncker the R^t Reverend Father in God Seth Ld Bishop of Sarum, our W^l[ellbeloved] Sr Samuel Morland K^{nt} & Bar^t, Sr Christopher Wren Kn^t, Our Surveyor Generall, Colonell Silius Titus, Dr John Pell, & Robt Hook Mt of Arts Greeting. Whereas the Sieur de St Pierre has humbly informed Us, that he hath found out the true Knowledge of the Longitude, and desires to be put on Tryall thereof; Wee having taken the same with Our consideration, and being willing to give all fitting encouragement to an Undertaking soe beneficiall to the Publick; and being alsoe well satisfyed with your abilities and Knowledge in matter of this nature; We have thought fit, and accordingly hereby doe constitute & appoint you, or any four of you, to meet together with what convenient speed you may, in order to the making, or causing to be made, and giving the said Sieur de St Pierre certaine Observations which he hath desired may be given to him

The height of two fixt Stars from the Horizon in exact degrees and minutes; the lesse these Stars decline from the Æquator or Æquinoctiall soe much the better;

The Elevation of ye Pole exactly in Degrees & minutes;

The Height of the Superiour & inferiour Limbs of the Moone, in Degrees and Minutes;

Whether these two Starrs be East or West in respect of the Place, where these Observations are to be made.

These Observations to be made exactly & at the same time, And the names of the said Stars, the yeare, Moneth & day, that the said Observation shall have been made, to be given likewise; And you are to call to your assistance such Persons, as you shall think fit: And Our pleasure is that when you have had sufficient Tryall of his Skill in this matter of finding out the True Longitude from such Observation, as you shall have made and given him, that you make Report thereof together with your opinions thereupon, how farre it may be practicable and usefull to the publick. Given at the Court at Whitehall the 15th day of December 1674 in ye 26th yeare of our Reign

By his maj^{ys} Comd

J Williamson" [7]

Samuel Morland arranged a meeting in Colonel Titus's house "in the Square of South-hampton buildings" on Friday 12 February 1675. In Pell's manuscript memorandum of this and ensuing events,[8] only Morland himself, Titus, Pell, and Hooke are recorded as being present; but he notes that "Mr Flamsteed of Darby was by consent of all fower, taken in as Assistant". John Flamsteed, a guest of Moore's whom the latter introduced to the meeting, had already earned himself a reputation as a skilful self-taught astronomer. He had no hesitation in firmly censuring the Sieur's proposal but the committee, considering the interests of the Sieur's patronage at Court, "desired to have him furnished according to his demands"; whereupon, Flamsteed undertook to supply the required data.

On the following (Ash) Wednesday, Flamsteed brought Pell details of two observations containing the information which the Sieur had requested, dated 23 February 1672 and 12 November 1673. On the next day, at a meeting of the Royal Society at Gresham College, he gave Pell three further papers incorporating an account of how the altitudes were obtained and his own adverse views on the proposal. These were all delivered without delay to the Sieur in the presence of a witness; but instead of being pleased that his request had been met, the Frenchman complained that they were out-dated and were calculations rather than observations.[9] Pell answered him to the effect that if he expected more modern data he would have to be patient until they were made, and that calculations could easily have been feigned for a recent date. News of these proceedings reached the King, who was surprised to learn that such great uncertainties still existed in the basic astronomical data required in navigation. He accordingly decided that Flamsteed himself, of whose talents he already knew through the intermediary of Moore, be made responsible for observing the heavens anew, with a view to examining and correcting not only the positions of the fixed stars but also those of the Moon and planets for the use of his seamen. Whereupon the following Warrant was issued from the Court at Whitehall on 4 March 1674/5 in Charles's name:

"Whereas, we have appointed our trusty and well-beloved John Flamsteed, master of arts, our astronomical observator, forthwith to apply himself with the most exact care and diligence to the rectifying the tables of the motions of the heavens, and the places of the fixed stars, so as to find out the so much-desired longitude of places for the perfecting the art of navigation. Our will and pleasure is, and we do hereby require and authorise you for the support and maintenance of the said John Flamsteed, of whose abilities in astronomy we have very good testimony, and are well satisfied, that from time to time you pay, or cause to be paid, unto him, the said John Flamsteed, or his assigns, the yearly salary or allowance of one hundred pounds per annum; the same to be charged and borne upon the quarter-books of the Office of the Ordnance, and paid to him quarterly, by even and equal portions, by the Treasurer of our said office, the first quarter to begin and be accompted from the feast of St. Michael the Archangel last past, and so to continue during our pleasure. And for so doing, this shall be as well unto you as to the Auditors of the Exchequer, for allowing the same, and all other our officers and ministers whom it may concern, a full and sufficient warrant.

Given at our Court at Whitehall, the 4th day of March, 1674–5.

By his Majesty's command,
J. Williamson.

To our right-trusty and well-beloved Counsellor,
Sir Thomas Chichely, Knt, Master of our Ordnance,
and to the Lieutenant-General of our Ordnance, and
the rest of the Officers of our Ordnance, now and
for the time being, and to all and every of them."[10]

Meanwhile, it would appear that the Sieur had not been satisfied with the observations that Flamsteed had communicated to him, since he was so persistent in troubling the King with requests for further data that Pell was officially instructed by the Secretary of State, Joseph Williamson, just over two months later to obtain a final judgment from the committee without further delay. Pell showed this letter to Flamsteed, Moore and Hooke and two days later, on 25 April 1675, Flamsteed hastily composed a letter to Pell for the information of the commissioners and Sir Joseph Williamson in which he emphasized the validity of his own observations, explaining what the Sieur's method must have been in order that the simultaneously observed values of two stellar and lunar altitudes, and the latitude of the unknown place of observation, were required for its solution. He also pointed out that substantial errors of almost $\frac{1}{4}°$ were inherent in contemporary lunar tables, while further uncertainties arose in estimating refraction and parallax. In a Latin letter bearing the same date but intended for the Sieur himself, Flamsteed stresses that the latter was not entitled to assume the Moon's celestial position, parallax, and angular semi-diameter to have been precisely established, nor exact measurements to have been made of refraction, which varies in accordance with the fluctuating state of the terrestrial atmosphere. This phenomenon alone could produce errors of at least 8′ in the Moon's position, corresponding to an error of 16 minutes (or 240 nautical miles) in longitude, which was enough to render the method impracticable for use at sea. He recognized that it would be impossible for a seaman to measure altitudes to within the required accuracy of ±1′ from the deck of a ship on the high seas. On account of the Sieur's evident lack of understanding of basic essentials such as the definition of altitude and the method of calculating the lunar parallax, Flamsteed was not inclined to credit him with having invented the method that he had been proposing, but suspected him of having borrowed it from his deceased countryman Morin, Longomontanus, or from some other neglected author.[11]

Very soon after writing these letters, Flamsteed demonstrated how the same data for 12 November 1673 that he had supplied to the Sieur de St Pierre ought to be applied to the calculation of the longitude difference from London of the unspecified place of observation in latitude 54° 00′ N. His calculations yielded the result that these observations had been made at a place 35 minutes East of London which, he points out, was over $44\frac{1}{3}$ minutes less than it ought to have been.[12] This remark enables us to identify the site as Hevelius's observatory at Danzig. The corresponding error in angular measurement is 11° 5′, which he attributes to the predicted lunar distance being at least 6′ less, and the Moon's calculated celestial latitude 3′ less, than they actually were at the time of observation. These discrepancies arose from his adoption of Tycho Brahe's erroneous star-positions and the existence of substantial errors of up to ±20′ in the lunar tables of Jeremiah Horrox on which Flamsteed had based these calculations, and for which he himself had re-computed the orbital parameters prior to their publication in John Wallis's edition of Horrox's posthumous writings (1673).

In order to improve the situation, nothing less than a complete revision of Tycho's star catalogue and a new lunar theory founded upon accurate measurements of the Moon's celestial position with respect to the re-determined stellar coordinates was to be achieved. And it was perfectly evident that if Flamsteed were to accept this enormous challenge by devoting his life to the fulfilment of such a laborious task, not only did he need the proper instruments with which to measure the positions of celestial bodies more accurately than any of his predecessors had done, but also a proper observatory to house them in. Moore had in fact been acting on his behalf by viewing locations within London where such an observatory might be established, and considered Chelsea and Hyde Park to be suitable areas for this purpose. However, the choice of site ultimately fell on Greenwich, which was recommended by Christopher Wren. A tower had formerly stood there which, for more than a century, had been used at different times as a residence, a prison, or latterly in the Civil War period as a castle. Not surprisingly, it is referred to as a castle in Charles's Warrant of 22 June 1675 authorizing the construction of the Royal Observatory:

"Whereas, in order to the finding out of the longitude of places for perfecting navigation and astronomy, we have resolved to build a small observatory within our park at Greenwich, upon the highest ground, at or near the place where the castle stood, with lodging-rooms for our astronomical observator and assistant, Our will and pleasure is, that according to such plot and design as shall be given you by our trusty and well-beloved Sir Christopher Wren, Knight, our surveyor-general of the place and scite of the said observatory, you cause the same to be fenced in, built and finished with all convenient speed, by such artificers and workmen as you shall appoint thereto, and that you give order unto our Treasurer of the Ordnance for the paying of such materials and workmen as shall be used and employed therein, out of such monies as shall come to your hands for old and decayed powder, which hath or shall be sold by our order of the 1st of January last, provided that the whole sum, so to be expended and paid, shall not exceed five hundred pounds; and our pleasure is, that all our officers and servants belonging to our said park be assisting to those that you shall appoint, for the doing thereof: and for so doing, this shall be to you, and to all others whom it may concern, a sufficient warrant.

Given at our Court at Whitehall, the 22nd day of June 1675, in the 27th year of our reign.

By his Majesty's command,
J. Williamson.

To our right-trusty and well-beloved Counsellor,
Sir Thomas Chichely, Knt, Master-General of our
Ordnance."[13]

The fact that many of the building materials were already at hand greatly facilitated the construction. More bricks were brought to Greenwich from Tilbury Fort, while some wood, iron and lead were obtained from a gatehouse demolished in the Tower of London. The foundations were laid on 10 August 1675 and the shell of the building completed before Christmas. In order to supervise the progress of this work, Flamsteed had moved in July from Sir Jonas Moore's house at the Tower to lodgings in Greenwich. He continued to make observations during the latter half of that year from the Queen's house, which now forms part of the imposing structure of the National Maritime Museum at the foot of the northern slope leading up to the Royal Observatory. His living-quarters in the Observatory which, on 10 July 1676, he began to inhabit, together with two servants Thomas Smith and Cuthbert Denton, consisted of a small dwelling on the ground floor of the building that has come to be called Flamsteed House. His astronomical observations were carried out from the Great Room directly above it from 19 September of that year. Thus did Britain's first scientific institution come to be established, for the clearly defined utilitarian purpose of providing the astronomical basis for a reliable solution of the urgent problem of oceanic navigation. John Flamsteed's appointment can scarcely be regarded as marking "the commencement of modern astronomy", nor can he be credited as having been "the first to apply the telescope and pendulum clock to astronomical observation", as his enthusiastic nineteenth-century biographer Francis Baily would have us believe;[14] however, in his capacity as First Astronomer Royal, he created a lasting tradition which E. Walter Maunder, in the concluding paragraph of his book *The Royal Observatory Greenwich* (London, 1900) published by the Religious Tracts Society, summarizes as follows:

> "Fundamentally, Greenwich Observatory was founded and has been maintained for distinctly practical purposes, chiefly for the improvement of the eminently practical science of navigation. Other enquiries relating to navigation, as, for intance, terrestrial magnetism and meteorology, have been added since. The pursuit of these objects has of necessity meant that the Observatory was equipped with powerful and accurate instruments, and the possession of these again has led to their use in fields which lay outside the domain of the purely utilitarian, fields from which the only harvest that could be reaped was that of the increase of our knowledge. So we have been led step by step from the mere desire to help the mariner to find his way across the trackless ocean, to the establishment of the secret law which rules the movements of every body of the universe, till at length we stand face to face with the mysteries of vast systems in the making, with the intimate structure of the stellar

universe, with the apparently aimless, causeless wanderings of vast suns in lightning flight; with problems we cannot solve, yet cannot cease from attempting, problems to which the only answer we can give is the confession of the magicians of Egypt—'This is the finger of God'."[15]

NOTES

1. The foregoing argument is expressed more fully in Forbes, 1971 c, where references are given to further sources.
2. It would appear to have been due to aberration, discovered by Bradley,1728.
3. McKie and Purver take opposing views.
4. B.M. Add. MS. Birch 4393, f. 37.
5. Kirk, p. 13.
6. These treatises were all published posthumously under Moore's name in 1681.
7. P.R.O., London: S.P. Dom. 44/334, pp. 27-28.
8. B.M. Add. MS. Birch 4394, ff. 26-108.
9. A memorandum of the foregoing events, in Flamsteed's own hand, is in R.G.O. MSS. P.R.O. Ref. 50, f. 262[r].
10. Baily, 1835a, pp. 111-112. Original in S.P. Dom. 44, p.10.
11. B.M. Add. MS. Birch 4393, ff. 96[r], 99[r,v], 101[r,v].
12. *Ibid.*, ff. 104[v], 104[r,v].
13. Baily, 1835a, p. 112. Original in S.P. Dom. 44, p.15.
14. *Ibid.*, p. xxv.
15. Maunder, p. 316.

2

Bricks without Straw

J OHN FLAMSTEED's first-known astronomical observation was of
a solar eclipse in his home-town of Derby on 12 September 1662,
when he was still only sixteen years of age. His interest in this
subject had been aroused by his reading of Johann Hevelius's
Mercurius sub sole visus published in Danzig that same year, and he had
just acquired the basis of his mathematical knowledge from a
borrowed copy of Sacrobosco's *Sphaerae*. His father, who was a
maltster, was soon to teach him the elements of arithmetic, how to
handle vulgar fractions, and how to solve problems of direct and
inverse proportion. Without further guidance, he made a cursory
study of several books on the art of constructing sundials and calculated a table of solar altitudes at all hours for the Equator, the
Tropic of Cancer, and the latitude of Derby (= 52° 55′ N.) with the
aid of a table of natural sines in Thomas Fale's *Horologiographia*.
He used the logarithmic tables in William Oughtred's *Trigonometrie*
to make further computations for the latitude of Derby, and adapted
the tables thus derived to a method of dialling described in Edmund
Gunter's *Description and Use of the Sector*.

Not long afterwards he cultivated the friendship of two local
gentlemen called George Linacre and William Lichford. From the
former he learnt much about the general appearances of stars and
constellations; and from the latter he learnt of the existence of a
reliable set of revised astronomical tables by Jeremiah Horrox in a
work by the well-known contemporary astrologer John Gadbury,
on which Lichford himself based his predictions of celestial phenomena. The cautious Flamsteed, however, was anxious to avoid
being linked with Gadbury or dubbed as an astrologer; thus he
bought Thomas Streete's *Astronomia Carolina* as an alternative and
more respectable basis of his own computations of planetary positions. Shortly before setting off on a journey to Ireland for health

reasons in August 1665, he wrote a "Mathematical Essay" on the construction and uses of a quadrant and ruler, containing tables for framing a quadrant to be used in the latitude of Derby. Soon after his return, about a month later, he presented this essay to Litchford, after adding an appendix showing the projection of a universal dial, and a catalogue of seventy stars. The equatorial and celestial co-ordinates were calculated for the year 1701 from values published in the Augsburg (1666) edition of Tycho Brahe's *Historia Coelestis* on the assumption that the annual precession of the equinoxes was 50″.

The direction that Flamsteed's interests were to take is evident from the nature of these youthful researches. His calculation from Streete's tables of the solar eclipse on 22 June 1666 so impressed Immanuel Halton of the neighbouring country estate of Wingfield Manor that this gentleman lent him a Latin copy of the 1647 edition of Noël Durret's "Richleian Tables", and subsequently the first volume of Giovanni Battista Riccioli's *Almagestum novum* (Bononiae, 1651) which contains a wealth of information on astronomical theories and methods from ancient times until the mid-seventeenth century. Riccioli's book was one of the most important literary sources of Flamsteed's knowledge. He read it through systematically during the following winter, transcribing some of the sections which particularly interested him. Among these is that Jesuit author's treatment of Aristarchus's method for estimating the relative distances of the Sun and Moon from the Earth by timing the instant at which the Moon's surface appears half-illuminated by the Sun's rays, which Flamsteed criticizes in his autobiographical account of his childhood years.[1]

His initial *telescopic* observations at Derby in the Spring of 1671 made him conscious too of the practical difficulty in estimating the apparent times of this dichotomy on account of the irregularities in the Moon's surface; thus he resolved to abandon that method of distance determination and look for another, which he was soon to discover in Johann Kepler's *Astronomia nova*. Here Kepler's concern had been to find the Mars–Sun distance from Tycho's observations of Mars's diurnal parallax and of the annual changes in the Sun's apparent angular diameter. Flamsteed, however, decided to invert this procedure by accepting the theories of these bodies' motions (derived by Kepler from Tycho's data) as published in *Ioanni Heckeri ephemerides motuum coelestium ab 1666 ad 1680* (Gedani, 1662), and seek the value of the Earth–Sun distance or the solar parallax. A daytime observation with his 3-foot radius wooden quadrant of the apparent angular separation between the Sun and Mars, when compared with the true (geocentric) distance interpolated from Hecker's ephemerides, yielded the sum of their parallaxes. Several nighttime observations with a telescope and micrometer eyepiece of Mars's distance from neighbouring fixed stars enabled its celestial motion to be quantitatively determined and afterwards eliminated to obtain that planet's parallax. Combining the results of these observations, he found that the solar parallax was not more than 10″. Similar measurements on three subsequent suitable occasions at Greenwich in 1676, 1679, and 1681 served merely to confirm this conclusion—a fact which was a tribute both to the consistency of the method and the accuracy of the observations.

The most significant feature of this method was Flamsteed's application of a screw micrometer eyepiece to his telescope which enabled him to measure the small angular distances between Mars and the stars used in this investigation to the hitherto unattainable accuracy of 1 or 2 seconds of arc. He was not the first to adopt such a technique. Adrian Auzout and Jean Picard in Paris had previously made systematic use of the former's own micrometer for similar measurements. He was, however, the first to use it to obtain his astonishingly close approximation to the value of the solar parallax, which one now accepts as being 8·80″. The particular instrument used was presented by Sir Jonas Moore to Flamsteed during his first visit to London in the summer of 1671. Sir Jonas himself had acquired it from Richard Towneley of Towneley in Lancashire, who had had it constructed by one of his tenants (referred to as an ingenious watchmaker) in accordance with a description of the original model made by its inventor William Gascoigne of Middleton, Yorkshire, in a manuscript letter to William Crabtree of Broughton near Manchester dated 1 November 1641, but incorporating some minor modifications of his own. This is why Flamsteed always alluded to it as 'Mr. Towneley's micrometer', rather than Gascoigne's micrometer. Robert Hooke published a description and drawing of it in the *Philosophical Transactions of the Royal Society* for 1667 after Towneley, in an earlier issue of the same periodical, had drawn attention to Gascoigne's priority over Auzout in this matter and explained how the micrometer could be applied to astronomical observations.

The success of this new approach to the determination of solar and planetary distances convinced Flamsteed of the great advantages to be obtained from equipping telescopes with micrometer eyepieces, and of applying these to quadrants or sextants in order to obtain more refined measurements of angular distances in the heavens lying outside the narrow field of a small-aperture refracting telescope. It also made him highly critical of contemporary observations made with plain sights. Thus, immediately prior to citing the results of these Derby observations in the *Phil. Trans.* for December 1672, he effectively rejects on this ground the entire value of Hevelius's lifelong labours to rectify the erroneous star positions in Tycho Brahe's *Historia Coelestis*. Hevelius felt obliged to defend himself in his *Machina Coelestis pars prior*, published in the following year, which induced a more aggressive attack by Robert Hooke in his *Animadversions on the First Part of the Machina Coelestis* (London, 1674) where, amongst other things, the latter maintains that the use of telescopic sights increased the accuracy of observation by a factor of ten.

The dispute is continued in a subsequent exchange of letters between Hevelius and Flamsteed. In a letter written about the end of 1676,[2] Hevelius rightly reprimands Flamsteed for attaching far too great a weight to a single observation made with the micrometer and regarding this as the norm for assessing the accuracy of data obtained by others. He lists five reasons why systematic errors could arise in taking the Sun's meridian altitudes with an astronomical sextant, though in his own case he was not prepared to concede an absolute inaccuracy in this quantity of more than $\pm\frac{1}{2}'$, and claimed that the internal consistency obtained from repeated measurements was of the order of $\pm 1''$. A further point which he makes in connection with their respective observations of Mars's position with respect to the stars in the Pleiades is that the discrepancies found may have arisen not in the act of observing but rather in the method of reduction which Flamsteed had employed. The latter had, for example, used Tycho's erroneous positions for the comparison stars, and worked upon the incorrect presupposition that Hecker's solar ephemerides—in effect, an extrapolation from Kepler's Rudolphine Tables which were themselves dependent upon Tycho Brahe's observations—contained no appreciable error.

Flamsteed was already well aware of the existence of differences of over $\pm 5'$ between Tycho's observations and his own, which he blamed largely on the latter's use of plain sights but also upon the uncertain effects of atmospheric refraction. He also knew that the unreliability of the value of the obliquity of the ecliptic was another factor affecting the values of the celestial longitudes and latitudes of stars calculated from Tycho's measurements of meridian altitudes with a large mural quadrant. In order to pursue the aim of rectifying the stellar coordinates (near the ecliptic and the Moon's orbit, particularly), Flamsteed realized that he required a mural quadrant equipped with telescopic sights, an accurate clock for timing the stars' meridian transits, and a reliable theory of atmospheric refraction. Moreover, he had to effect a ten-fold improvement in the predictive ability of contemporary solar ephemerides like those of Streete's and Hecker's to attain an accuracy of within $\pm 1'$.

Sir Jonas Moore, as Surveyor General of the Board of Ordnance, was responsible for ensuring that Flamsteed received his salary. But no money had been set aside to pay for the instruments themselves, so it was necessary to be as economical as possible. Thus, when Hooke assured him that he could construct a mural quadrant at low cost, Sir Jonas accepted his offer and commissioned him to undertake that work. This proved to be a most unfortunate decision, since the quadrant was so badly designed and constructed that Flamsteed soon found that the trouble and care which he was obliged to exercise in using it were not justified by the quality of the results obtained. He was somewhat more fortunate with a 6-foot 9-inch radius iron sextant with telescopic sights which was made to his own specifications and constructed under Sir Jonas's personal supervision and at his expense by Edward Silvester, the Master Smith at the Tower of London. He mounted this on its axis and semicircles in September 1676 and used it during the next three months to measure the angular distances outwith the meridian from the Sun, Moon, planets, and stars to other stars. These observations taught him that Frederigo Cesi's idea of revolving such an instrument in the plane of the declination circle by means of a crown wheel and handle geared to its serrated limb, which Flamsteed had adopted on the basis of Hooke's recommendation in the *Animadversions*, was impracticable, owing to the wearing down of the male screw attached to the end of the movable index. The calibration table relating the revolutions of that screw to the angular displacement of the sextant was therefore no longer reliable; thus the instrument had to be dismounted, and its limb divided diagonally in the conventional manner described by Hevelius and Tycho. Thereafter, the screw-errors were determined with reference to this scale. Flamsteed's mode of applying the telescope to such an instrument was to position the object-glass over the centre, or pivotal point, the eyepiece on the outer edge of the limb, and the focal plane with cross wires on the inner edge of the limb. The width of the limb was therefore determined by the magnifying power of the telescope: it was raised by a fraction of an inch above the plane of the instrument so that the plumb-line through the pivot was not obstructed. The zero point of the limb was found by inverting the quadrant and bisecting the difference between the two points where the plumb-line crossed the outer limb.

The impracticability of Hooke's quadrant for obtaining reliable meridian altitudes made Flamsteed decide to adapt his sextant to that purpose, after he employed it in December 1676 to measure the meridian altitudes of stars culminating close to the zenith (where refraction could be ignored) and to obtain the greatest and least altitudes of the Pole star. This enabled him to estimate instrumental errors and obtain an approximate value for the latitude of the Royal Observatory (viz. $51° 28' 10''$), which is $28''\cdot2$ less than the modern accepted value. However, the lack of stability in the mounting and the fact that the instrument was not properly balanced, resulted in an internal inconsistency of nearly $\pm1'$ being incurred in these relative observations, which he regarded as too great to justify their being published.

Flamsteed's interest in the accurate timing of the meridian transits dates back to early studies at Derby in 1665, which were aimed initially at disproving the existence of real differences in the time-intervals between successive transits of the Sun. Instead, they led him to the correct conclusion that the length of the 'natural' (apparent solar) day really does vary periodically owing to the combined effect of the eccentricity of the Earth's annual orbit and its obliquity to the plane of the celestial equator. His demonstration of these irregularities was composed in Latin seven years later and published under the title "De Inaequalitate Dierum Solarium Dissertatio Astronomico" as an appendix to John Wallis's edition of *Jeremiae Horroccii Opera Posthuma* (Londini, 1673). Its fundamental tenets are that the Earth's diurnal rotation is uniform, an assumption which is implicit in the conversion of terrestrial longitude differences from time-units into angular measure; and that the net inequality

Bricks without Straw

in the length of the solar day should be identified with the difference between the Earth's mean and true right ascensions. These were nevertheless but 'rational conjectures', and Flamsteed knew that practical trials with a good pendulum clock were required to establish their validity. Thus he resolved to make the accurate experimental determination of this Equation of Time the object of one of his first investigations at the Royal Observatory, and encouraged Richard Towneley, with whom he was already in regular correspondence, to conduct similar experiments. Pendulum clocks were still not common in Britain at this time, and had to be specially made.

The most important specifications for such timekeepers were that they should be as long as possible and driven by a very heavy weight, in order to be insensitive to changes in temperature and humidity. Their rates would be less seriously affected by dust clogging with the oil in the pivot holes, and Flamsteed considered that a greater regularity in their oscillations would also be attained by sloping the pallets in a manner recently introduced by the London clockmaker Thomas Tompion. Thus Tompion was commissioned to make two clocks with 13-foot long pendulums in accordance with these requirements, and duly delivered them to the Royal Observatory on 7 July 1676 as Flamsteed was just about to take up residence there. Along with the sextant, a 52-foot focal length object-glass, and some books, they were a gift to him from Sir Jonas Moore.

Flamsteed employed two independent methods of regulating these clocks. One was to note the times of equal altitude measurements of the Sun or a bright star near the zenith, and interpolate the clock-time corresponding to local apparent noon. The instrument which he initially used for this purpose was his own 3-foot radius wooden quadrant with plain sights which he had brought with him from Derby. He dispensed with this in June 1678 after he had divided another 40-inch radius quadrant made by Hooke, procured from him by Sir Jonas Moore from the Royal Society at the beginning of the year, and attached it to a rotatable pivot. However, only a month after the latter's death on 27 August 1679, he was ordered to return it together with a 5-inch radius quadrant with a screw-limb and another with two telescopes on it that had been acquired at the same time. He was therefore induced to fit up a 50-inch radius quadrant of his own, with which he could sight consistently upon a star to within $\pm\frac{1}{2}'$ and thereby estimate his clock-errors to about

±3 seconds. An alternative method, which was recommended to him by Towneley in October 1677, was to time the transit of stars of nearly equal altitudes (or declinations) past a meridian wall; but since it was no easy matter to guarantee that the wall lay precisely in the north–south direction, he had little faith in that method and finally abandoned it less than two years later.

In order to regulate a third pendulum clock (which Towneley's servant had made for him) to beat out sidereal time Flamsteed, early in 1677, made a 'wall telescope' consisting of two 6-foot focal length convex lenses in brass cells attached to an iron ruler fitted to one of the walls, and used it to observe the times of successive transits of Sirius at noon. He noted the sidereal times at the moments when the Sun's altitude was taken with the Royal Society's, or subsequently with his own 50-inch radius quadrant. The difference between the corrected values of the corresponding sidereal and mean solar times was then the Equation of Time. His initial comparisons between these experimentally derived figures and those previously found by theoretical demonstration yielded differences of between $+0^s\cdot45$ and $-1^s\cdot12$ per day, and therefore appeared to confirm the assumption that the Earth's axial rotation with respect to the stars is isochronous. Thus, by the end of 1677, this 'first principle of astronomy' had been raised above the level of a mere postulate, and verified experimentally to within the accuracy attainable with the Tompion clocks.

In the meantime, however, Flamsteed had not been sitting idle. His lack of a properly designed mural quadrant had certainly temporarily prevented him from carrying out the conventional observations for determining the right ascensions and declination of the Sun and stars with the desired degree of precision (viz. ±1'). On the other hand, it had been responsible for his developing an alternative method for obtaining these data; one which he then thought to be far better in principle, but which he later came to regard as much less convenient in practice. He employed it primarily for investigating the Sun's apparent motion—in effect, the Earth's motion—since a reliable solar theory was "the only firm basis of all astronomy" and necessary for the rectification of planetary coordinates. It involved measuring with his sextant the Sun's angular distances from Venus or Jupiter by day, and either of these planets' distances from a bright zodiacal star by night, from which he deduced the apparent separation in celestial longitude between the Sun and that star with perhaps

32

twice the accuracy attainable by timing the difference in their transits with this Towneley pendulum clock and observing their meridian altitudes with the sextant itself. One might be inclined to suppose that the use of Mars as an intermediary in finding the solar parallax was responsible, either consciously or subconsciously, for Flamsteed's decision to use these other planets as celestial stepping-stones for obtaining the Sun's separation in longitude from the fixed stars. In fact, the idea was borrowed from Tycho Brahe's *Astronomia instauratae progymnasmata* (Francofurti, 1610).

The principles of this method, as adapted by Flamsteed, can be inferred from a tract entitled "De motu Solis correcto Commentarius" preserved among his unpublished manuscripts at the Royal Greenwich Observatory.[3] In Fig. 1, let S, V, X, denote the positions of the Sun, Venus, and a star respectively on the celestial sphere. SK, VK, XK are longitude circles through each of these bodies and the pole of the ecliptic K. SAB is an arc of the ecliptic (through S) meeting KV produced in A and KX produced in B; thus AV, BX are the celestial latitudes of
Venus and the star respectively. The problem which Flamsteed had to solve was: given SV, VX by measurements with the sextant find, with the aid of Thomas Streete's Caroline Tables (1661) for the Sun and Venus, and Albertus Curtius's edition of Tycho Brahe's *Historia Coelestis* (1666) for the star, the net separation in longitude SB between the Sun and the star. The analysis was essentially as follows:

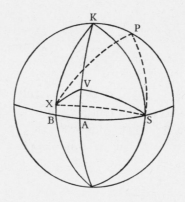

Fig. 1

In the right-angled spherical triangle SVA:

$$\cos SA = \cos SV \sec AV.$$

Since SV was observed and AV interpolated for the day-time observation from Streete's solar tables, SA is found from this special case of the standard cosine formula of spherical trigonometry. In the right-angled spherical triangle VKX:

$$\cos \widehat{VKX} = \frac{\cos VK - \cos VK \cos KX}{\sin VX \sin KX}$$

33

c

But $\widehat{VKX} = \widehat{AKB} = AB$ and $VK = 90° - AV$, $KX = 90° - BX$; hence

$$\cos AB = \frac{\cos VX - \sin AV \sin BX}{\cos AV \cos BX}.$$

Since VX is observed, AV interpolated for the night-time observation from Streete's ephemeris for Venus, and BX extracted straight from Tycho's star catalogue, AB is found from this general case of the cosine formula. Hence $SB = SA + AB$.

Using his approximate value for the latitude of the Royal Observatory ($\varphi = 51° 28' 10''$) and the approximate altitude of the star (h) at the time of the second observation, Flamsteed obtained the latter's declination (δ) from the formula $\delta = \varphi + (90° - h)$ and hence its north polar distance $90° - \delta = 180° - \varphi - h = 128° 31' 50'' - h$. Combining the cosine formula for the spherical triangles SBX and SPX, one immediately obtains

$$\cos SB \cos BX = \cos SX = \cos PX \cos PS$$
$$+ \sin PX \sin PS \cos \widehat{XPS}$$

or

$$\cos \widehat{XPS} = \frac{\cos SB \cos BX - \sin BX \sin \delta_\odot}{\cos BX \cos \delta_\odot},$$

where the star's celestial latitude $BX = 90° - PX$ was known from Tycho's tables reprinted in Riccioli's *Almagest* and the Sun's declination $\delta_\odot = 90° - PS$ was easily deducible from the known values of its celestial longitude ($\lambda_\odot$) and the obliquity of the ecliptic ($\varepsilon = 23° 29'$) by means of the transformation formula

$$\sin \delta_\odot = \sin \lambda_\odot \sin \varepsilon.$$

In order to obtain the value of the Sun's true celestial position SB, corrections had first to be applied for the change of several minutes of arc in Venus's longitude during the interval between the day and night observations and the variable effects of atmospheric refraction. All the quantities on the right-hand side of the last equation now being known, the difference in right ascension between the Sun and the star ($= \widehat{XPS}$) could finally be calculated. The effect of precession of the equinoxes on the star's longitude during the interval between the epoch of Tycho's catalogue (viz. 1600.00) and that of Flamsteed's own observations, and the influence of Tycho's having

34

adopted too large a value for the obliquity of the ecliptic when computing the latitudes of stars, had both to be eliminated before the Sun's right ascensions at different times of year could be tabulated.

It was merely in order to apply these refraction corrections that approximate values for the altitudes of the Sun and Venus at the times of observations were required; the catalogued stellar co-ordinates being provisionally assumed free from this particular source of uncertainty. Knowing these altitudes to the nearest degree from his sextant observations, Flamsteed first obtained from his tables the corresponding values of refraction in altitude; and calculated its components in latitude and longitude from the circumstances of the observations (viz. the latitude, date, and time) with the aid of auxiliary tables provided by Streete for this purpose. The advantages of the method were that neither did the values for the obliquity of the ecliptic nor the latitude of the observatory, nor the precise longitudes of the zodiacal stars, require to be explicitly known to effect the first phase of the calculation; namely, the determination of the Sun's longitude *difference* from each star. Flamsteed was also able to minimize the influences of refraction by selecting times for his relative observations at which the altitudes of the three celestial bodies concerned were as high and as nearly equal as possible. He considered that errors of up to $\pm 2'$ in the values of refraction would produce no perceptible contraction in the measured angular distance, especially since in the region close to the horizon where the phenomenon itself is most pronounced the close parallelism of the latitude circles greatly reduces any such tendency.

Flamsteed was by this time well aware of the urgent need to develop both an empirical and a theoretical basis of refraction. At Derby he had measured the angular diameters of the heavenly bodies in the neighbourhood of the horizon with Towneley's micrometer fixed to the eyepiece of a 7-foot focus telescope, and found ample evidence of apparent variations that could only have been caused by the changing state of the terrestrial atmosphere. Vapours, fogs, mists, clouds, the Aurora Borealis, etc., all appeared to constitute direct observational evidence for the increase in the density of the air closer to the Earth's surface. His observation that a mercury column decreased by 0·14 inch when carried to a height of 142 feet up the steeple of All Saints Church in London, implied a proportional decrease of 1/209 in the density of the air, but he correctly surmised that in the case of measurements made at different heights above sea-level an equal decrease in atmospheric pressure need not correspond to the same difference in altitude. It was in the hope of obtaining an experimental verification of this fact that he had written to Towneley towards the end of 1673 requesting him to make independent measurements of the height of a mercury column on a local landmark known as Pendle Hill.[4] A complicating factor, of whose existence he had even then been well aware, was that observations of the mercury column made at the *same* place but at different times could also be affected by temperature changes in the atmosphere; and it was his recognition of the fact that refraction increases with temperature which was responsible for his acceptance of Tycho's figures of 34', 33' and 30' for the mean horizontal refraction of the Sun, Moon, and stars respectively. Towneley and he thus began exchanging the results of such observations, and each drew up a set of rules by which they could be used to forecast the weather. Sir Jonas Moore, intrigued by this, had got Flamsteed to construct two such weather-glasses or barometers for him with which he had been able to amuse the King, Charles II, and the young Duke of York—a trivial but perhaps not insignificant occurrence since it preceded the issue of the Royal Warrant appointing Flamsteed as the King's 'Astronomical Observator'.

Flamsteed's comparison in 1674 between his own results on refraction and those Tycho cited by Riccioli showed that the latter were too big near the horizon and that their proportional decrease was too slow above 30° altitude. The refraction table published in Tycho's *Astronomiae instauratae progymnasmata* from which they were taken was of an empirical nature, and was strictly applicable only to the site of the Uraniborg observatory. The reason for Giovanni Cassini's assertion that the effect was detectable right up to the zenith as his own observations indicated, was provided by "Mr Edmond Halley, an ingenious young man from Oxford"[5] who began corresponding with Flamsteed early in 1675 on that and other astronomical subjects, and who was soon to be lending him some valuable assistance with certain of his early observations at Greenwich. The fundamental premise in Halley's trigonometrical demonstration, communicated by Flamsteed to Towneley in a letter dated 29 April 1675, invoked the principle that each light ray suffered a single refraction somewhere in the upper atmosphere and then proceeded in a straight line to the observer.

His own observations during June 1678 using Hooke's 40-inch

radius quadrant with telescopic sights convinced him that, at very low altitudes of about $\frac{1}{3}°$, morning refraction could exceed evening refraction—a result that he interpreted as evidence for vapours emitted from the Earth at night being dispersed by the heat of the Sun's rays during the course of the day. Thus apart from vapours, fogs, mists, clouds, the Aurora Borealis, etc., the variability of the horizontal refractions from day to day seemed also to constitute direct observational proof of the increase in density of the atmosphere closer to the Earth's surface; thereby implying multiple or continuous refraction, and that the ray would follow a curved path. For this reason he rejected the theory of refraction used by Cassini in the Malvasian ephemerides and subsequently those of Kepler, Jean Picard, and Halley's ex-university friend Charles Boucher after he discovered that they had also been constructed on the same false principle of a homogeneous sphere of vapours about $2\frac{1}{2}$ miles high. The refraction tables published by Thomas Streete and Vincent Wing were merely copies of Tycho's empirical tables, and likewise stopped at only 20° altitude instead of being extrapolated in the direction of the zenith.

When Newton later produced a table of refractions based upon a theory involving a uniform decrease of air density with height Flamsteed was quick to accept it as a satisfactory alternative; but the former discarded it when he became aware that it implied a refracting power at the top of the atmosphere which was as big as that at the base. During the early months of 1695, Newton constructed a new table of refraction using observational data collected in June 1678, February and April 1681, which Flamsteed had supplied. This time, however, the former postulated that the density of the air was proportional to the compression produced by the action of the inverse-square law of universal gravitation upon the particles from which it was composed, an alternative to which Flamsteed was again ready to assent, particularly when he saw how minute were the differences between the corresponding values in the two tables. All Flamsteed's corrections to measurements of angular distances and celestial altitudes are based upon the revised figures, which were judged correct to within 1″ for all altitudes above 10° and to within 3″ for those between 3° and 10°. Regrettably, he did not follow Newton's suggestion that thermometric and barometric readings ought to accompany every astronomical observation.[6]

Flamsteed's faith in the Copernican viewpoint would seem to have stemmed not from philosophical considerations but from another attempt to test the influence of refraction on the accuracy of his observations. Following a procedure which had previously been adopted by Hevelius, he selected seven zodiacal stars from a total of about a hundred that he had been using as standards of comparison for determining the positions of fainter stars and the planets; his intention being to see whether the sum of their calculated differences in right ascension would make up one complete revolution, as ought to be the case if their zenith distances and relative angular separations measured with the 6-foot 9-inch radius sextant were accurately known. He found that the total contraction in right ascension due to refraction was 5′ 24″, which meant that the sum of the right ascensions ought to have been less than 360° by this amount, whereas in fact it was observed to be only 38″ less. Such was his confidence in his observational technique and method of reduction that he attributed the difference of 4′ 46″ to the existence of a physical cause producing a *dilation*, which could be "no other than the parallax of the earth's orb".[7]

This was a rather premature conclusion with which Flamsteed tried to impress Seth Ward, the Bishop of Salisbury, when he wrote to him on 31 January 1680 to acquaint him with the progress of his studies in the hope of winning his favour and support at Court. The death of his friend and patron Sir Jonas Moore just over five months earlier had left him in a precarious financial position which dashed his hopes of receiving funds from the King for the construction of a mural quadrant, and obliged him to begin teaching young gentlemen whom he could also on occasion use as unpaid assistants. A manuscript preserved among his papers at the Royal Greenwich Observatory entitled "A list of my pupils' names and employments, as far as my memory will serve me"[8] and covering the period 1666 to 1709, cites about 140 persons including several members of the English aristocracy, captains of vessels, and trainees for service with the East India Company. Flamsteed told Ward that he had found evidence of an acceleration in the motion of Saturn, and of a smaller deceleration in that of Jupiter. Hecker's ephemerides for Mars, which he had come to accept as the best then in existence, required little correction, but the tabulated positions of Venus and Mercury were often found to err by ±15′ or more and the Moon's longitudes were almost as irregular.

The principles on which Flamsteed had been basing his calculations of these planetary and lunar positions prior to this time are illustrated in Fig. 2, which retains the notation of Fig. 1 but substitutes a second zodiacal star Y in place of the Sun. In general, Y does not lie exactly in the ecliptic; thus C is introduced to denote the point at which the longitude circle KY intersects the ecliptic. V may be taken to represent Venus itself or any of the other planets. In observing the Moon, the telescopic sighting was generally made on the point of the limb nearest to the star from which its distance was being taken,

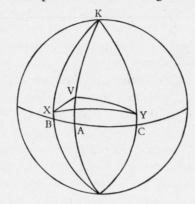

Fig. 2

and a reduction made to the centre of its disc by applying the appropriate value of the semi-diameter obtained from previous measurements with the micrometer.

In △ KXY:

$$\cos KY = \cos KX \cos XY + \sin KX \sin XY \cos \widehat{KXY}$$

Since

$$KX = 90° - BX \quad \text{and} \quad KY = 90° - BY,$$

$$\cos \widehat{KXY} = \frac{\sin CY - \sin BX \cos XY}{\cos BX \sin XY}$$

where the stars' latitudes BX, CY were reduced from Flamsteed's own observations or taken from Tycho's tables, and XY was measured with the sextant. Thus $\widehat{KXY}$ was known. Similarly, from a consideration of △ VXY, $\widehat{VXY}$ was known. Thus $\widehat{KXV} = \widehat{KXY} - \widehat{VXY}$ was found.

In △ KXV:

$$\cos KV = \cos KX \cos XV + \sin KX \sin XV \cos \widehat{KXV}$$

or

$$\sin AV = \sin BX \cos XV + \cos BX \sin XV \cos \widehat{KXV}$$

Since XV was measured, the latitude AV was derived from this equation. Also

$$\frac{\sin \widehat{XKV}}{\sin XV} = \frac{\sin \widehat{KXV}}{\sin VK}$$

or

$$\sin BA = \sin \widehat{BKA} = \sin \widehat{XKV} = \frac{\sin XV \sin \widehat{KXV}}{\cos AV}$$

from which, since all the quantities on the right-hand side are known, the difference in celestial longitude from the star X (=BA) was deduced.

Thus, by means of three angular distances taken nearly simultaneously with his sextant, and a prior knowledge of the stars' latitudes, the planet's or Moons' latitude and longitude difference from either one of these stars could be found and compared with the corresponding quantities interpolated from Hecker's ephemerides to yield the errors. What was still lacking was the position of the equinoctial points on the ecliptic from which the longitudes of all these heavenly bodies could be measured, but this was unobtainable by such relative measurements. Flamsteed therefore took the longitude of the bright star of Aries from Tycho's catalogue, and reduced it to his own epoch (1679) by allowing 50″ for the annual precession. He preferred his own value for its celestial latitude deduced from the meridional altitude measured with his sextant. This was the point to which all his observed places of the Sun, Moon, planets, and stars were now referred. Any error in the position of the equinoxes incurred by the uncertainty in the position of α Arietis could be compensated by noting two times when the Sun's meridional distances from the zenith—hence its declinations— observed near either equinox were the same, and its true distance from the latter point found by interpolation from tables of the Sun's mean apparent motion.

The new solar tables constructed by Flamsteed in 1679 on the basis of the foregoing considerations were a distinct improvement on those that had been published six years earlier in the appendix to Wallis's edition of Horrox's *Opera Posthuma*. The eccentricity of the annual orbit was still based on the micrometric measurements begun in Derby of the Sun's angular diameter at aphelion and perihelion; however, the aphelion position was now 1° less, the greatest inequality 48 seconds more, and the resultant accuracy in calculating the Sun's celestial longitude was seldom inferior to ±1′. The fact that the Sun's mean motion required no revision facilitated the task of incorporating these amendments into the earlier ephemerides. Although he had established the Sun's apparent motion in the plane

40

of the ecliptic, Flamsteed had still to determine the obliquity of that orbit with respect to the celestial equator. This quantity is necessary not only for evaluating one of the two periodic components constituting the Equation of Time, but also for the transformations of horizontal and equatorial coordinates into ecliptic ones, and conversely; thus any imprecision in its amount tends to affect both the accuracy of the solar theory and the celestial longitudes and latitudes of stars calculated from the meridian transit observations which Flamsteed still intended to make once he obtained a suitable mural quadrant. Should it prove subject to a gradual variation with epoch, one would expect such changes to have been reflected in the values of the latitudes contained in the star catalogues of Ptolemy, Ulugh Beigh, Tycho Brahe, and others.

Accordingly, Flamsteed made a study of literary sources describing the methods which had been used for determining the obliquity, with a view to establishing both its constancy and magnitude. Riccioli's book sufficed for a knowledge of Greek and Roman attempts. His information on Arabic and Persian observations was gained from a letter written to him on 14 August 1681 by Edward Bernard, the Savilian Professor of Astronomy in Oxford and a noted oriental scholar;[9] from John Greaves' edition of *Shaw Cholgii Persae Astronomica* (Londini, 1652); and from some manuscripts of the late John Selden that he was able to borrow out of the Bodleian Library, Oxford. He further examined the results obtained by Edward Wright, Philip Lansberg, Albert Lineman, Giovanni Riccioli, Giovanni Cassini, Gottefrid Wendelin and Johann Hevelius, taking account of refractions and of incorrect allowances that had been made for the solar parallax. His conclusions were that the celestial latitudes and declinations of the stars observed by the Ancients were scarcely ever reliable to within $\pm 1°$, and that the obliquity had always remained constant at $23° 29'$.

This well-considered result, and Flamsteed's firm belief in Copernican astronomy, were fundamental to "The Doctrine of the Sphere" composed by him in 1680 and printed in Sir Jonas Moore's posthumously published book *A New Systeme of the Mathematicks* (London, 1681). Part 1 of this lengthy dissertation was an account of a stereographic projection of the heavens in which both the diurnal and annual appearances were represented using the Pythagorean assumptions of the Earth's axial and orbital motions. Part 2 was concerned with the application of this doctrine to solar and lunar eclipses. The novelty of this work lay in his choosing the *invariable* plane of the ecliptic as the plane of projection, coupled with his adoption of the ancient postulates upon which Copernicus had founded his controversial system of planetary motion.

The idea of representing celestial phenomena in this manner seems to have originated in a conversation which Flamsteed had with Halley early in 1676 in which the latter asked if he were able to determine the magnitude and phases of a solar eclipse without having recourse to calculation. It was the appearance of sunspots in the summer of that year which stimulated him to consider how the Earth would appear if viewed from the Sun, and how the diurnal phenomena would appear if projected on it. The stereographic method which he developed yielded the Sun's zenith distance, the parallactic angle, the celestial longitude difference between the nonagesimal degree and the Sun, and the altitude of the nonagesimal degree at any time of day; thus it enabled all the circumstances of the eclipse to be graphically determined. The diurnal and nocturnal arches, equator, latitude circles, and hour circles could all be drawn and divided into equal degrees using ruler and compasses only, since the most important property of this kind of representation is that circles on the celestial sphere are projected into circles in the plane of projection. He was able to show how a knowledge of the Sun's longitude and the obliquity of the ecliptic provided by the researches described above was sufficient for the graphical determination of that body's right ascensions and declinations at any specified time; and how the ecliptic coordinates of a star could also be plotted on this projection.

Flamsteed appended to "The Doctrine of the Sphere" the solar tables which he had just calculated in 1679 and a revised version of the lunar tables that Wallis had published with Horrox's *Opera Posthuma* in 1673; together with further tables for the calculation of eclipses. He not only supervised the printing of these writings, but also that of Perkins's treatise on navigation which the latter had just managed to complete before his death; thereby discharging this obligation to the memory of his two recently departed friends in the Spring of 1681.

By this time, however, Fate had channelled his astronomical endeavours into yet another direction. The appearance of a comet in November of the previous year had excited widespread interest in the scientific world, and several of his close acquaintances such as

Bernard and Halley in Oxford, James Crompton in Cambridge, and Fellows of the Royal Society had been expressing an interest in learning his thoughts or knowing his observations on it. Besides this comet's spectacular appearance, its orbit was worth investigating to see whether it conformed to Hevelius's general opinion in his *Prodromus Cometicus* (Gedani, 1666) "that without the annual Motion of the Earth, no rational Account can be given of any Comet, but that all is involved with perplexities, and deform'd by Absurdities". Was its path a conic section with the Sun at one focus?, as Kepler had shown to be the case for planets. Or was it rectilinear?, as the latter had believed and as Hevelius himself had decided to favour by the time he composed his *Cometographia* (Gedani, 1668).

It was his reading of Regiomontanus's and Apianus's accounts of comets in the second volume of Riccioli's *Almagest* which led Flamsteed initially to accept the notion that comets travel along a great circle arc in the heavens, and to reject the currently held view that they move in straight lines. Through Crompton, he was led into a correspondence with Newton which, as Francis Baily (1835a) has remarked, was "interesting as showing us the state of the science at that period; and the good understanding that *then* existed between Newton and Flamsteed".[10] The latter's analysis of cometary motion is a development of the same method as he had used to determine the solar orbit, which enabled him to determine the comet's ecliptic coordinates, the length and direction of its tail, and the node of its path. Observations made at Greenwich between 12 December 1680 and 5 February 1681 supplemented by others from Abraham Hill in Canterbury, Jean Charles Gallet and Antonio Collio in Rome, and G. D. Cassini in Paris, appeared to indicate that the comet experienced a slight attraction towards the plane of the ecliptic which became less as it receded farther from the Sun; only at remote distances from that luminary was the motion along a great circle arc. Other complicating features were actually due to errors incurred by Flamsteed in his computations. He attempted to explain these data by postulating the existence of a solar vortex motion and a magnetic force of attraction and repulsion, but Newton preferred to re-formulate the theory in terms of an attraction only coupled with a directional property that caused the comet to orbit round the Sun and be centrally attracted towards it. It would appear from the latter's correspondence with Flamsteed that about April 1681 he decided in favour of gravitation as the causal agency, partly because he could not bring himself to believe that very hot bodies like the Sun could possess a magnetic virtue.[11]

By now Flamsteed realized that the only hope of obtaining the mural arch which he so much desired lay in making it himself at his own expense, with the help of an unskilled workman. It was begun in August 1681, and completed before the end of that year; but its 140° wooden limb was so badly warped that for a long time he was discouraged from dividing it. Finally, in 1683, he did so. The instrument was fixed on a meridional wall, and had been specially designed to "take in all the stars that passed the meridian betwixt the pole and the south intersection of the meridian and horizon",[12] including the Pole star itself. It was used by Flamsteed for measuring the zenith distances of the Sun and planets until the autumn of 1686, but the absolute values of this quantity were unreliable on account of the lack of rigidity in its structure and its tendency of the wall attachments to become loose. Nevertheless, he had enough faith in the 'intermutual distances' (i.e. the relative differences in zenith distance between one heavenly body and another) taken with it to use these in conjunction with his sextant observations to investigate the major inequality in the Sun's motion, and to prepare a catalogue of about 130 stars.

The year 1687, remembered in the annals of science as the year of publication of Newton's *Principia mathematica Philosophiae naturalis*, marks the culmination of the initial phase of Flamsteed's work at Greenwich. During the twelve years that had elapsed since his appointment, the First Astronomer Royal had continued the micrometric measurements of solar, lunar, and planetary diameters begun at Derby and, using new tubes with 16 and 8-foot focus objective lenses, made frequent telescopic observations of lunar occultations and small angular distances in the heavens before his telescopical sextant was fitted for use. Thereafter, he made about 20,000 observations with that instrument. He used these data

(i) to examine the inequalities in the motions of the Sun, Moon, and planets; in the course of which he established the existence of the long period inequality of Jupiter and Saturn;

(ii) to establish the basis of his new star catalogue; and

(iii) to investigate the motion of the Comet of 1680.

He succeeded in experimentally establishing his earlier theoretical demonstration of the Equation of Time, provided a firmer observational basis for the theory of atmospheric refraction, and critically

reviewed previous attempts to establish the value of the obliquity of
the ecliptic. His belief in the constancy of the latter quantity and in
the validity of Copernicus's postulates of the Earth's diurnal and
annual motions led him to conceive a new graphical representation
of all these celestial appearances, and to apply it to the calculation
of the circumstances of eclipses, and lunar occultations. Its value as a
method of rectifying the solar and lunar ephemerides diminished
in the late eighteenth century when the errors in these were reduced
to only a few seconds of arc but, in spite of what the French
historian of astronomy, Joseph Delambre, says to the contrary,[13]
Flamsteed deserves credit for originality in devising it.

Deeper insight into these and other aspects of Flamsteed's activity
prior to 1687 may be gained from a study of his extensive correspondence, notably with Henry Oldenburg, John Collins, Sir
Jonas Moore, Giovanni Cassini, Richard Towneley, Edmond Halley,
Johann Hevelius, Isaac Newton and William Molyneux, which is
presently being collated with a view to publication[14]. His simultaneous preoccupation with the evaluation of the solar parallax, the
theory of cometary motion, the theory of telescope optics, the
application of telescopic sights to astronomical instruments, the
errors in ancient observations, the stereographic ecliptic projection
of the celestial sphere, the physical constitution of heavenly bodies,
the evaluation of the precessional constant, and the theory of
astronomical refraction, is revealed in the volume of his thirty-nine
lectures on astronomy delivered at Gresham College, London,
between April 1681 and November 1684—a hitherto untapped
source of information on his early intellectual development that has
just become available in printed form[15], as a contribution to the
tercentenary of the institution that he did so much to promote.

NOTES

1. Baily, 1835a, p. 22.
2. Hevelius to Flamsteed, 23 December 1676 [Old Style], in Hevelius, 1685,
   pp. 79–89.
3. R.G.O. MSS., P.R.O. Ref. 44.
4. Flamsteed to Towneley, 25 November 1673, R.S. MSS. F.1.2.
5. Flamsteed to Towneley, 29 April 1675, R.S. MSS. F.1.5.
6. Newton to Flamsteed, 24 October 1694, in Scott (Ed.), p. 36.
7. Flamsteed to Ward, 31 January 1679/80, in Baily, 1835a, p. 20.

8. R.G.O. MSS. P.R.O. Ref. 15 (B), ff. 162^v, 166^{r,v}, 167^r.
9. Published in *Phil. Trans. R. Soc.*, **14** (1684), 721–725.
10. Baily, 1835a, p. 51.
11. This correspondence, which was conducted through the intermediary of James Crompton in Cambridge, is contained in the third volume of Turnbull (Ed.), 1961.
12. Baily, 1835a, p. 54.
13. Delambre, 1827, pp. 99–100.
14. Mansell Ltd. has undertaken to publish Flamsteed's Correspondence in three volumes: Vol. 1, 1666–84; Vol. 2, 1685–1701; Vol. 3, 1702–19, plus supplementary correspondence to 1730.
15. Forbes, 1975.

3
The Establishment of a Tradition

THE second phase of Flamsteed's work at the Royal Observatory begins with his application of a new 7-foot radius mural arch similar in design to the first but much stronger, constructed by his skilled assistant Abraham Sharp between August 1688 and October 1689. The cost (of over £120) was again borne by the Astronomer Royal himself, but fortunately his financial situation had just been improved by an inheritance from his late father's estate. It did not take long for him to establish that there was nothing to choose in accuracy between his former method of calculating, from sextant observations of intermutual and zenith distances, the differences in right ascension and declination between the Moon or one of the planets and bright zodiacal stars, and the more direct method of observing the differences in the times of their meridian transit and in their zenith distances. The latter was much more convenient in practice, requiring the assistance of only one other person to read the clock and take notes instead of two helpers to manipulate and take sightings with the sextant. Moreover, the observer was now sheltered within the building rather than exposed to the weather on the roof. The narrow aperture of only $1\frac{1}{2}$ feet in width through which the sky could be seen, by reducing stray light, enabled stars of the 7th magnitude to be detected with the naked eye. The accurate collimation of the telescopic sights was a further guarantee that his observations would now possess a much greater degree of internal consistency than Tycho had attained with his large mural quadrant.

In order to investigate the deviation of the plane of Sharp's instrument from that of the meridian, Flamsteed compared the times of the Sun's transits over the meridional thread in the telescope index with those deduced from equal altitude observations of the Sun

47

before and after local apparent noon. He thereby established that the transit on the tropic occurred 38 seconds earlier than it should have done, indicating that the alignment of the instrument was then $9\frac{1}{2}'$ west of the true meridian; and in 1690 he prepared a correction table for other times of the year as well.

Great care had also been taken by Sharp and Flamsteed in dividing this quadrant with a wooden beam compass, and neither suspected at the time that there might have been any tendency for the beam to yield and thereby affect the accuracy of the markings.

The conditions were now right for collecting new and reliable observations of star positions, and Flamsteed worked hard at this for over two years before drawing up his 'tables of transits' containing the raw data arranged under the names of constellations, of the transit times, meridional zenith distances, north polar distances, magnitude class, and right ascensions standardized on that of some principal star in each constellation. From these the true equatorial and ecliptic coordinates of every star could be found, and charts of the constellation drawn.

On 10 August 1691, Newton wrote to Flamsteed advising him to publish a catalogue of all those stars down to the sixth magnitude which had been observed by earlier and contemporary astronomers, followed by an appendix of others observed by himself alone. In replying to this request some six months later[1], Flamsteed gave specific instances of inconsistencies which he had found in Tycho Brahe's catalogues and Johannes Bayer's celestial charts, and explained why in his opinion it would be useless to make such a distinction. He asked Newton whether he ought now to cease observing before his survey of the heavens was complete, in order to make time for preparing his results for the press; or rather, continue while he still had the health and vigour to prosecute this task. This query was purely rhetorical, since Flamsteed began soon afterwards to engage, at his own expense, a number of helpers to expedite the computation of the extensive data which he had been steadily assimilating. Luke Leigh, working in Derby, calculated the celestial longitudes and latitudes of the stars from the observed right ascensions and distances from the North Pole. Seven manuscript volumes preserved among Flamsteed's papers at the Royal Greenwich Observatory are testimony to the steady progress which he made with this chore. More stellar and planetary reductions were made by a relative of Christopher Wren named James Hodgson (who, in 1702, married Flamsteed's niece Anna), Thomas Witty, Luke Leigh, and William Bossley of Bakewell. Bossley concentrated his efforts primarily upon Jupiter and Saturn, and attempted to improve the theories of these planets' motions. Thomas Weston "an ingenious but sickly youth" began to draw the figures of the constellations in accordance with the descriptions of Ptolemy.

Just before the turn of the century, Flamsteed composed a report on the state of his work throughout the entire period of his appointment as Astronomical Observator at Greenwich, of which the following (written in the third person) is an extract:

"The observations, taken with this instrument [the second mural arc] in 10 years past, since it was built, are fair described in two folio volumes; and are in number above five and twenty thousand. From the solar observations (a part of them) in the year 1694, he [Flamsteed] derived solar tables; and from them, has lately made tables of the sun's declinations, for the use of our sailors, that will serve them 100 years.

The number of the fixed stars, visible with the naked eye, observed by him with this instrument, he accounts about 3000: of which he had, since then, determined the right ascensions, distances from the pole, longitudes and latitudes of above fifteen hundred; adding the variations of both, answering to one degree change of longitude. Whereby the right ascensions and declinations of them may be got, to any time past or to come; which will be of great use to our sailors. And this has been done with the help of only one servant, and a calculator hired at a great distance in the country [viz. in Derby], at his own charge, in five years past, since he made his new solar tables.

In the mean time, his observations have been continued, as occasion required; and persons of known ability and skill [Newton] have been furnished with some hundreds of measures of refraction, and places of the superior planets and moon, derived both from them and the best tables extant; in order to the rectifying the tables of refractions, and determining the motion of the planets, the moon's especially; of which we hope, in a short time, to have, by this means, such tables as shall represent her places in the heavens, very nearly.

The eclipses and configurations of Jupiter's satellites have also been continually observed, as the weather permitted. And new tables of their motions will be made, which will be of great use to our sailors for correcting the longitudes of known coasts, or finding those of unknown [but only after more experience and further observations had been gathered].

49

There remain:

1. The right ascensions and distances of about 1500 fixed stars, to be determined from the observations already taken; with their longitudes and latitudes: and the above-mentioned variation, to be determined from them, in order to complete the catalogue.

2. The places of the moon to be calculated, from the observations taken with the sextant, betwixt the years 1676 and 1689; which require persons of more than ordinary skill and patience to compute them, being in number betwixt four and five hundred.

3. As also her places, to be derived from observations made betwixt 1689 and 1699, above 300.

4. Maps of the constellations to be made anew (or copied from those already made), to be printed with the catalogue of the fixed stars.

5. The volumes of observations to be copied for the press (that the originals may be preserved, to be recurred unto in case of any doubt, mis-print, or suspected error), together with tables already made, to be used with them."[2]

Flamsteed estimates that this work could be completed in two or three years with the help of six or more skilled assistants, but in about twelve years with his existing helpers. By then, the planetary coordinates and theories based upon them would be published in order that sailors could correct their sea-charts and improve their knowledge of position at sea. Natural philosophy would also benefit. Money for bigger and better instruments and for the printing of these various works would, however, be required; which possibly explains why he is still prepared to acknowledge the king as "The greatest and best patron of the useful arts of peace, as well as of war", despite earlier complaints concerning the inadequacy of his salary.

There is no indication as to whether Flamsteed intended this paper for any particular personage (e.g. Sir John Worden) or merely to allay the criticisms of those who condemned him for his failure to publish his observations, but he does say in one of the fourteen notes which he appended to it a year or so later that it had been seen by some of his friends. In the interim the above-mentioned assistants had calculated about 300 lunar places both from the mural arc observations and Horrox's tables; the places of Jupiter, Saturn, and Mars observed at the oppositions and quadratures of the Sun between 1689 and 1700; new tables for the systems of Saturn, Jupiter, and the Moon; and were engaged upon rectifying 381 star positions in eight constellations, which left only two more to be done. We know from a letter, written on 10 May 1700 to John Lowthorpe,[3] that Flamsteed had paid Newton a visit one week previously to acquaint him with the current state of progress, informing him that the calculations for the zodiacal stars and southern constellations were complete, and that work was continuing on 400 stars in the northern constellations. Further calculations were being made from 50 observations of Venus, 40 of Jupiter, 30 of Saturn, and over 200 of the Moon at the oppositions of the Sun and quadratures, all made during the previous decade with the mural arc. Newton had openly acknowledged that his theory, both of cometary motion and that of the Moon, had been based solely upon the observational data which he had received from Flamsteed. On the subject of the projected star catalogue, Flamsteed told Newton that the observations themselves ought first to be published, and charts engraved to accompany the catalogue after suitable measures had been taken to ensure that map-makers would not pirate that work and transcribe it into their globes.

John Wallis, who was less well informed about this progress, wrote that in the public interest he "would be glad to hear that your observations are in the press, that so great a treasure be not lost, of which we are in great danger in case you should die before they be printed"[4]. Flamsteed assured him that his catalogue would shortly contain some 2200 star places, and that only five constellations remained to be examined. He had found over 100 places of Mars from his own observations between 1671 and 1701, and a similar number for Saturn and Jupiter between 1689 and 1701, all computed using the new places of the fixed stars. Two hundred places of the Moon between 1689 and 1695 had also been determined from a small star catalogue that had been imparted to Newton. Thus there was no danger of his "long and painful labours"[5] being wasted. After the completion of the catalogue, good and skilful assistance would still be required to calculate the planetary coordinates from the measurements taken with the sextant between 1676 and 1689, and with the new meridian arc, outwith the times of quadrature and opposition. Tables would also be appended in order to facilitate the calculation of those planetary positions. Among these were new tables of the Moon's mean motion and anomaly, horary motions, parallaxes, and semi-diameters, calculated by Flamsteed and James Hodgson, and modified tables of the apogee and node. He expresses his bitterness at the prejudiced views of theorists who had no

51

conception of the calculations involved in reducing astronomical observations, which were embodied in over a dozen handsome quarto volumes (besides what his Derby assistant had filled) and a couple of folios of collections and synopses of the constellations.

One man who could render the kind of help that Flamsteed needed was his former assistant Abraham Sharp from Bradford who, after having worked at Greenwich for almost six years, had left the Royal Observatory on 4 November 1690 to teach mathematics in London. Three months later, he gave this up in favour of a more advantageous job at Portsmouth, possibly working for a shipwright on the manufacture of nautical instruments—a position which he held for almost three years before returning to London. But after the death of his elder brother the Rev. Thomas Sharp in the autumn of 1693 he felt obliged to manage the family property at Little Horton, near Bradford, and look after his brother's widow whose young son was about to begin studying for the medical profession. For this reason he decided to return to that estate just before Easter of the following year instead of applying for the post of Mathematical Master at Christ's Hospital which had become vacant as a result of Edward Pagett's resignation. He would very likely have been the successful candidate, since he had the support of Flamsteed himself, Edmond Halley, John Parsons, Euclid Speidell and John Colson.

Sharp would appear to have initiated his lengthy and scientifically interesting correspondence with Flamsteed on 2 February 1702, just when the latter was in desperate need of qualified assistance. After replying to several points in Sharp's letter, Flamsteed acquaints him with the state of his star catalogue at that time enclosing a sample copied by Weston of how his data would be set out in the star catalogue. He appears to have taken cognisance of an earlier complaint of Sharp's against Hevelius by introducing in the last two columns the amounts by which the right ascensions and north polar distances vary for each 1° increase in celestial longitude. Only four constellations now remained to be done, two of which were nearly complete. Flamsteed had been comparing the 'altogether ungeometrical' maps of Hevelius and Bayer, and the old editions of Ptolemy's star catalogue by Lucas Gauricus, Copernicus, George Trapezuntius, Thomas Hyde and Sebastian Münster. John Caswell had consulted a manuscript in Oxford on his behalf to eliminate the remaining inconsistencies. He had already copied Tycho's and Hevelius's catalogues in accordance with his own preferred arrangement, and intended to annex that of Ulugh Beigh after similarly copying it from Hyde's edition. Nevertheless, it would require "a great deal of time to finish these figures; more, to complete the constellations wanting, to copy the observations for the press, to calculate the moon's and planets' places from those made with the sextant".[6]

Newton visited the Royal Observatory on 12 April 1704 only a few months after his election as President of the Royal Society, to ascertain what further progress Flamsteed and his poorly paid assistants had been making with this laborious work. He was shown the observation books, the star catalogues of Flamsteed himself, Tycho, and Hevelius, the star charts, and the engravings on which P. van Sommer "an excellent draughtsman, but in years"[7] was now also engaged. The time seemed ripe to seek financial support for publication, and Flamsteed allowed Newton to intercede publicly on his behalf with Queen Anne's husband George, Prince of Denmark, who, being already aware of the former's work, immediately consented to sponsor it. The plan of the envisaged *Historia Coelestis Britannica*, giving Flamsteed's estimate of the number of pages that his books of observations and catalogues might occupy in print—these totalled 1450 in all—was accidentally brought to the notice of the Royal Society at their anniversary meeting in November 1704, with the result that a committee was appointed to bring this document to the attention of the Prince, who confirmed his willingness to provide the necessary funds and proposed that Francis Robartes, Sir Christopher Wren, Dr John Arbuthnot, and any other gentlemen whom the Royal Society might deem competent, should inspect Flamsteed's papers and judge when they were fit for the press. These referees co-opted the assistance of Francis Aston, David Gregory and Newton, and Newton arranged a dinner party for them on 19 December to which Flamsteed brought some specimens of his work. Another meeting was held nine days later and subsequent talks between Newton and Flamsteed had to be held before a reliable costing and report could be drawn up on 23 January 1705 for the Prince's consideration. In it, the referees submitted as their opinion that this was the most extensive and comprehensive set of data that had ever been made, and necessary for the perfection of astronomy and navigation. The proposed publication was to cost £683, with an additional £180 for the further calculation and printing of the lunar, planetary, and cometary places, of which 600 out of the total of 2000 were already computed.

After this had been duly approved, Flamsteed took up Sharp's earlier offer of assistance and sent him books and instructions which enabled him to compute the Moon's observed places from the sextant measurements in the manner already described in Chapter 2. The zenith distance and parallactic angle were calculated from the ecliptic coordinates knowing the obliquity of the ecliptic and the latitude of the observatory. Flamsteed's own corrections for refraction and the parallax in altitude deduced from the former, and the corresponding parallaxes in longitude and latitude from the latter, enabled the Moon's true places to be found. Hecker's lunar ephemerides were not applicable after the year 1680, thus those of Andreas Argoli for the years 1671 to 1700 based on Tycho's lunar places were used in their stead. Allowances for the Moon's semi-diameter were based upon Flamsteed's early micrometric observations. These lunar computations, which Sharp completed in September 1705, were so laborious that even if he, as an experienced calculator, worked without interruption for a full eight-hour day, he could reduce only between six and eight lunar positions. The consequent danger of errors being committed in the process was revealed by the considerable number of discrepancies between the results obtained by Sharp and those calculated independently at Greenwich by Thomas Witty. In October 1705, Flamsteed sent to Sharp, Witty's places of Saturn, Jupiter, and Mars, uncorrected for refraction and parallax, to compare with his own calculations.

The "Articles of Agreement" between the publisher Awnsham Churchill, Flamsteed, and the referees, were finalized on 17 November 1705.[8] It would appear that Flamsteed then agreed to the printing of 400 copies of his projected two-volume work for which he would promptly supply the manuscript and read the proofs, and was granted free access to the press at all times with the right to stop it when that number of copies had been printed. The publisher promised to print twenty sheets per month and to have no further interest in the work after it had been completed, while the referees were responsible for disposing of the Prince's grant. Flamsteed was to be paid £125 on delivery of the complete manuscript of each volume. There were, however, significant differences between the draft agreement which Flamsteed wanted the referees to sign, and that which the latter persuaded him to sign. Under the terms of the contract, he legally bound himself to prepare and deliver to the referees "with all convenient speed . . . fair and correct copies"[9] of his star catalogue, and to have the latter published in the *first* volume. He was, however, unable to obtain any guarantee from them that they would give him the custody of the printed work. The fact that Flamsteed did put his signature to this agreement is confirmed by his written complaint to Sharp only a few days later:

> "Sir Isaac Newton has at last forced me to enter into articles for printing my works with a bookseller, very disadvantageous to myself."[10]

The crux of the bitter arguments which afterwards ensued between Newton and Flamsteed would seem to be the former's insistence on these conditions being honoured by Flamsteed coupled with the latter's refusal to abide by them, particularly after he saw that the printer was not keeping to his promised work-schedule. He also registered his strong objection to Awnsham Churchill being highly paid for his services, while he—the author—got little recompense for his own labours and financial outlays, but this was a matter for the referees only to decide. By the following March, Flamsteed had delivered his first observation book and 100 sheets of copy to Newton for printing, which were checked by Gregory whose lack of information about the values of the micrometer screw errors caused him to think that numerous errors had been incurred. Flamsteed dispelled this illusion by submitting a copy of his own table for correcting these along with his second and third observation books containing his notes to 10 September 1689.

On 15 April 1707, Newton and Gregory visited Flamsteed at Greenwich in order to find out how the work on the second volume and catalogue was progressing. They offered him a paper to sign, pledging to deliver to the referees a fortnight hence 174 sheets of the second volume (containing observations to the end of 1705) plus a corrected and completed copy of the star catalogue, of which an imperfect copy had been deposited with Newton under seal over a year earlier. He was reminded of the condition that these materials would have to be submitted before any of the money promised to him would be paid, and told that nothing more would be printed until he did so. Flamsteed refused to sign, since he knew he would be unable to satisfy the time-clause, but promised to proceed as quickly as he could. Yet almost another year passed before he finally handed over to the referees a few additional sheets for volume 1, the mural arc observations and a copy of the star catalogue which was still not complete. He was then given back the imperfect catalogue and asked to insert the stellar magnitudes into it, which he did, and duly returned it to Newton. His later assertions that he had to wait 'some months' before receiving his first and only instalment of £125 would seem to be exaggerated, since this payment was settled within four weeks.

Meanwhile, the threat that the press would be stopped pending the receipt of the mural arc observations and star catalogue had been put into effect. Flamsteed regarded this as sheer obstructiveness, since the printing was already proceeding at a much slower rate than it ought to have been. The referees' determination to adhere to the first "Article of Agreement", that the star catalogue should be published in volume I, probably accounts for their action, and why they became so impatient with Flamsteed's delay in bringing this part of his work to perfection. Thus they resolved on 13 July 1708

> "that if Mr Flamsteed do not take care that the Press be well corrected & go on with dispatch, another Corrector be imployed".[11]

In Flamsteed's own eyes, however, Newton had been personally responsible for hindering the publication of his work. He expresses his resentment at the tone of this resolution in a letter to Christopher Wren, for whom he always retained a high regard, and stresses his own desire to see everything completed. Despite the referees' intention of now proceeding with the printing, the death of the benefactor (Prince George) on 28 October 1708 occasioned a complete stoppage of the press and gave Flamsteed some welcome breathing space to get more of his data ready for publication. Besides the reduction of over 3000 star positions, these included calculations by his assistants Abraham Riley and Joseph Crosthwait of the planetary observations made at Greenwich since 1676, which numbered almost 1000 in all.

Flamsteed's satisfaction with this situation was short-lived, for he was officially informed in December 1710 of Queen Anne's intention to exercise a tight control over his professional affairs through the intermediary of a Board of Visitors to the Royal Observatory, consisting of a president, vice-president, and others from the Council of the Royal Society. These 'constant visitors' were to see that the existing instruments were in a serviceable condition, and authorized to purchase whatever others may be required. Flamsteed was commanded to give them an account of what observations he had made, and undertake whatever others that they may prescribe. The Queen's Warrant actually reads as follows:

> "To our trusty and well-beloved the President of the Royal Society for the time being.
>
> Anne R.
>
> Trusty and well-beloved, we greet you well. Whereas we have been given to understand that it would contribute very much to the improvement of Astronomy and Navigation, if we should appoint constant Visitors to our Royal Observatory at Greenwich, with sufficient powers for the due execution of that trust, we have therefore thought fit, in consideration of the great learning, experience, and other necessary qualifications of our Royal Society, to constitute and appoint you, the President, and in your absence the Vice-President of our Royal Society for the time being, together with such others as the Council of our said Royal Society shall think fit to join with you, to be constant Visitors of our said Royal Observatory at Greenwich: authorising and requiring you to demand of our Astronomer and Keeper of our said Observatory, for the time being, to deliver to you within six months after every year shall be elapsed, a true and fair copy of the annual observations he shall have made. And our further Will and Pleasure is that you do likewise, from time to time, order and direct our said Astronomer and Keeper of our said Royal Observatory to make such astronomical observations as you in your judgement shall think proper. And that you do survey and inspect our instruments in our said Observatory; and as often as you shall find any of them defective that you do inform the principal Officers of our Ordnance thereof; that so the said instrument may be either exchanged or repaired.
>
> And so we bid you farewell.
>
> Given at our Court at St. James's, the 12th day of December, 1710, in the ninth year of our reign
> By Her Majesty's Command
> H. St. John"[12]

After he had had time to swallow his indignation at what he considered to be "another piece of Sir I.N.'s ingenuity," Flamsteed decided against submitting a petition to the Queen and to place his faith in the influence of his own friends at Court who included the Lord Chamberlain, the Duke of Bolton, Robert Walpole, Sir Paul Methuen, Lord Torrington (Mr Newport). The Queen's physician Dr John Arbuthnot, in his capacity as a referee, soon pressed him to supply the complete copy of his star catalogue since the imperfect copy re-deposited under seal to Newton did not contain six of the northern constellations nor the computed celestial longitudes and latitudes of the stars. Failure to do so in the immediate future meant

57

that Halley, who had already made numerous changes to Ptolemy's names of stars adopted by Flamsteed to make these consistent with the nomenclature of his own catalogue of southern stars (cf. Chapter 4), would be appointed comparer and editor of the entire work, even if it meant his having to make do with Flamsteed's incomplete and unreliable catalogue. The very fact that Arbuthnot knew what was lacking in this catalogue was an indication that its seal had been broken, which led Flamsteed to accuse Newton personally of a gross breach of faith. The fairness of this censure has been a matter for debate, since Newton's undertaking not to break the seal had been given *before* the catalogue had been returned to Flamsteed in March 1708, and not after it had been returned by the latter (presumably re-sealed) shortly afterwards. Moreover, it would seem that Newton had the Queen's own express permission to do so. Whatever the moral and legal justifications for that action, Arbuthnot's ultimatum triggered off the following immediate indignant response from Flamsteed:

> "I have now spent 35 years in composing and work of my catalogue, which may, in time, be published for the use of her Majesty's subjects, and ingenious men all the world over. I have endured long and painful distempers by my night watches and my day labors. I have spent a large sum of money above my appointment, out of my own estate, to complete my catalogue, and finish my astronomical works under my hands . . . tell me ingenuously and sincerely, were you in my circumstances, and had been at all my labor, charge, and trouble, would you like to have your labors surreptitiously forced out of your hands, conveyed into the hands of your declared, profligate enemies, printed without your consent, and spoiled, as mine are, in the impression? Would you suffer your enemies to make themselves judges of what they really understand not? Would you not withdraw your copy out of their hands, trust no more in theirs, and publish your own works rather than at your own expense, than see them spoiled, and yourself laughed at, for suffering it?"[13]

Notwithstanding this outburst, Halley pressed on with his edition of Flamsteed's work, backed by the powerful influence of Newton, Arbuthnot and the rest of the referees. Newton even gave Halley the lunar observations based on Flamsteed's sextant observations which, in 1694, he had promised not to communicate to any other person. Doubtlessly, he was now under the impression that the agreement signed by Flamsteed in 1705 superseded this earlier understanding. To the latter, however, it seemed as if insult were being added to injury. When he became aware of the liberties being taken by Halley in editing the imperfect catalogue, he flatly refused to allow his completed copy to be subjected to similar treatment and decided to print it, and the observations from which it had been derived, at his own expense. He even began legal proceedings against Newton for refusing to return the 175 pages of mural arc measurements, since he firmly believed that the referees had no authority nor right to keep these papers following Prince George's death. There was a very real difficulty in deciding on whether they should be regarded as private or public property, especially since Flamsteed had by now spent about £2000 of his own money upon instruments, observing assistants, and calculators. Thus, after Halley's edition of Flamsteed's work had been published and distributed to public libraries and eminent individuals at home and abroad, to certain members of the British aristocracy, to promoters in the Government, and to the referees, as well as to the author himself, Newton decided to return the manuscripts. Sir Robert Walpole, as First Lord of the Treasury, arranged for Flamsteed to receive the remaining 300 printed copies, of which he retained 388 pages, and burnt the title-page and preface, the pirated star catalogue, and 120 pages extracted from the later observations—in all, about a quarter of what had been printed under the auspices of the referees.

In September 1716, Flamsteed began writing the English version of the Prolegomena (or Preface) to his own edition of the star catalogue, in which he takes a rapid survey of the progress of astronomy, concluding with a description of his own instruments and methods of observation. By the time of his death on 31 December 1719, this task had been completed and the planets' places derived from the mural arc observations reduced ready for publication, those up to July 1717 having been printed. The major difficulties facing Joseph Crosthwait in Greenwich and Abraham Sharp in Horton, Lancashire, who assumed responsibility for completing this literary monument to Flamsteed's memory, were those of finding a suitable person to latinize the Preface and reliable engravers for the charts of the constellations.

The translation of the Prolegomena presented a problem since few understood its subject-matter. It was begun by James Pound from Crosthwait's transcription of Flamsteed's original English version, continued by a grammar-master of the Christ's Hospital Mathematical School at Greenwich whose knowledge both of Latin and of astronomy proved inadequate, and completed by a presbyterian minister, the Reverend Anderson of St James's. The proofs were checked by William Whiston, formerly Newton's successor in the Lucasian Chair of Geometry at Cambridge, who, after his dismissal from that post for Arian heresies, was earning a living as a mathematics teacher in London. The complete absence of any reference to the circumstances under which the work was to appear has given rise to the speculation that part of what Flamsteed originally wrote has been suppressed; but the questions of when, why, and by whom remain unanswered.

Sharp himself, at the request of Flamsteed's widow, began engraving maps of the zodiacal constellations; but this work was subsequently transferred to a London firm of engravers in charge of a Mr Nutting, which had been strongly recommended by Sir James Thornhill. The two assistants primarily employed on this work were called Vandergucht and Vertue, but their prices were so high that Crosthwait journeyed to Holland and commissioned two Dutch engravers to undertake the same work at a lower fee. When the quality of their work did not measure up to expectations, Margaret Flamsteed temporarily stopped it on account of its great expense; but it was subsequently resumed and completed with the help of Vandergucht and "a young man in the Minories by Tower Hill"[14] in London.

Notwithstanding all these difficulties and delays, Flamsteed's *Historia Coelestis Britannica* was finally published in 1725. The first volume contains extracts from the observations of William Gascoigne and William Crabtree from 1638 to 1643, followed by Flamsteed's own Derby observations to 1675, and his subsequent Greenwich observations with the sextant from 1675 to 1689. It comprises various subsidiary tables to be used in calculations, and the right ascensions, north polar distances, celestial longitudes and latitudes of the Moon and planets deduced from the sextant observations. The second volume contains the mural arc observations from 1689 to 1719, to which are appended another collection of useful tables and a similar list of the lunar and planetary places computed from these data. The third volume begins with the Prolegomena, followed by copies of all the catalogues of the fixed stars that had been executed by earlier astronomers; these being succeeded in turn by Flamsteed's own catalogue exhibiting the equatorial and ecliptic coordinates of 2935 stars reduced to the beginning of the year 1689, together with the annual variations in right ascension and north polar distance arising from the precession of the equinoctial points. At the end, there are various auxiliary tables designed to reduce the labour of calculation.

This work has been praised by William Cudworth as occupying the same place in practical astronomy as Newton's *Principia* in the theoretical part. The star catalogue was described by Baily in a memoir to the Astronomical Society of London as "the proudest production (considering the period at which it was made) of the Royal Observatory at Greenwich".[15] This bracketed qualification is nevertheless extremely significant since, as Baily goes on to stress, the clock-rate was irregular because pendula had not yet been provided with a temperature compensation, analytical methods had not been developed, the phenomena of aberation and nutation (cf. Chapter 5) were still unknown, identifications of stars were quite often erroneous, refractions uncertain, and temperatures and barometric pressure at the times of observation were never recorded. By modern standards, therefore, Flamsteed's methods of reduction were very imperfect; yet, as the critical French historian of astronomy Joseph Delambre also pointed out, the little fruit that astronomy was able to pluck from the enormous amount of labour which went into the preparation of these volumes is not the fault of the observer, but rather that of the instruments which he had at his disposal.[16] Sir George Airy was of the opinion that in point of detail in observing methods, system in the presentation of the data, and exhaustiveness in the extraction of information from them, Flamsteed's work could shame any other collection of observations in Britain or any other country.

NOTES

1. Flamsteed to Newton, 24 February 1691/2, in Turnbull (Ed.), 1961, pp. 199–203.
2. Baily, 1835a, pp. 188, 189.
3. An extract is published in Scott (Ed.), pp. 331–332. The full letter is in R.G.O. MSS., P.R.O. Ref. 33, ff. 185^r–187^r.
4. Wallis to Flamsteed, 3 June 1701, in Scott (Ed.), pp. 365–366.
5. Flamsteed to Wallis, 24 June 1701, *ibid.*, p. 368.
6. Flamsteed to Sharp, 6 February 1702, in Baily, 1835a, p. 201.
7. Baily, 1835a, p. 65.

8. Scott (Ed.), pp. 454–459.
9. *Ibid.*, p. 455.
10. Flamsteed to Sharp, 20 November 1705, in Baily, 1835a, p. 256.
11. Scott (Ed.), p. 524.
12. Baily, 1835a, pp. 90–91.
13. Flamsteed to Arbuthnott, 9 April 1711, *ibid.*, p. 284.
14. J. Crosthwait to Sharp, 18 May 1724, *ibid.*, p. 356.
15. Baily, 1831, p. 131.
16. Delambre, 1827, p. 103.

4

The Price of Progress

AN important aspect of Flamsteed's work which has merely been mentioned but not discussed in the preceding chapter was his interest in the theory of the Moon's motion. This had been aroused at an early stage in his career, as a result of his careful study early in 1672 of the correspondence between the two North-country astronomers William Crabtree and William Gascoigne.[1] He had acquired these twenty or so letters (dating from 1640 to 1642) "by the singular favour"[2] of Richard Towneley, whose uncle (Christopher) had obtained them from the former's widow in Broughton near Manchester. Crabtree, writing to Gascoigne on 21 June 1642, cited seven precepts for determining the true celestial longitude of the Moon which he had found among the papers of his deceased friend the Reverend Jeremiah Horrox. Horrox had left no description of the theory itself, but Crabtree was helped in his reconstruction by rough diagrams drawn on loose papers which he drew more neatly for Gascoigne's benefit. This is reproduced from the original manuscript preserved among Flamsteed's papers, on the following page (cf. Fig. 3).[3] Crabtree's own interpretation of Horrox's lunar theory, which Flamsteed transcribed together with the precepts and diagram on which it had been based, is essentially as follows.

S denotes the Sun, and T the Earth. The larger circle is the Earth's annual orbit around the Sun, the smaller circle is the Moon's monthly orbit around the Earth. A, P are the positions of the apogee and perigee respectively; C, O the Moon's points of conjunction and opposition with the Sun. M is the Moon's mean apogee, E the centre and ET the eccentricity of the Moon's orbit.

In the positions numbered 1, 3, 5, 7 the Moon's true and mean apogees coincide; but as the Earth leaves positions 3 and 7, the true apogee A differs from the mean M by the angle $\overset{\frown}{\text{ATM}}$. This angle,

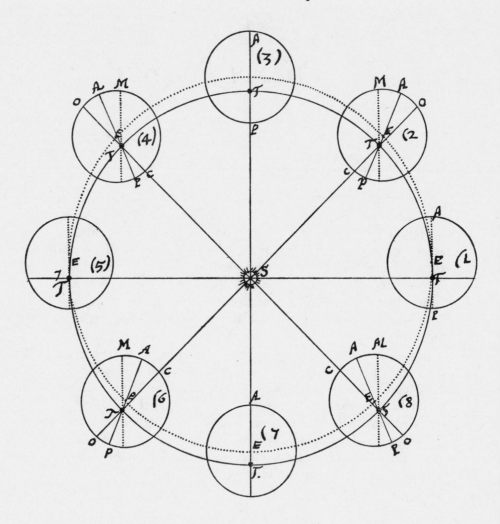

Fig. 3

or equation of the Moon's apogee, must be added to the mean between 3 and 5 or between 7 and 1 in order to obtain the true apogee; but between 5 and 7 or 1 and 3 it must be subtracted. This rule is evident from Fig. 3. The Sun appears to attract either the Moon's apogee or perigee depending on which is the nearer, the analogy being that of a loadstone with its magnetic virtue concentrated at its two poles. Thus in positions 1 and 5, where both A and P are at equal distances from S, or in 3 and 7, where the lines AS, PS are coincident, the net effect is zero. The value of the eccentricity is least in 1 and 5, and greatest in 3 and 7.

All but the last of Horrox's precepts could be understood from this description and the fact of the Moon's elliptical orbit around the Earth; but Crabtree was unable to demonstrate the seventh and last precept concerned with the inequality in the Moon's motion known as the variation of the eccentricity.[4] The following geometrical construction for it was supplied for the first time by Flamsteed in his epilogue to the "Novae Theoriae Lunaris a Jerem Horroccio primum adinventae, et postea in emendatiorem formam redactae ..." published in John Wallis's edition of *Jeremiae Horocii ... Opera Posthuma*; viz. *Astronomia Kepleriana, defensa & promota*: etc. (Londini, 1673).

Let ASPZ (Fig. 4) be the Moon's orbit, C its centre, and Cb its mean eccentricity. With b as centre, describe a circle of radius bF equal to half the difference of the greatest and least eccentricity,

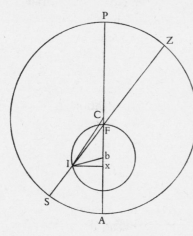

Fig. 4

Through F, draw the syzygiacal line SIZ cutting the circumference of this circle in I. Join Ib, and draw Ix perpendicular to AP. Then $I\hat{b}x = 2\ b\hat{F}I$, the angular separation of the apogee from the Sun; while Cx is the true eccentricity.

By thus coupling the libratory motion of the apsidal line AP with a variable eccentricity, Horrox (and subsequently Flamsteed) united the two principal lunar inequalities; namely, the equation of the centre and the evection. In his first set of lunar tables, mentioned above, Flamsteed followed Horrox in identifying the annual equation in the Moon's mean motion with the Equation of Time, thereby reducing the tables for finding the corrections of the Moon's longitude to two only. He was, however, aware of the inaccuracy of this substitution, which created a false distinction with the Equation of Time as applied to the Sun, and

D

restored both of these quantities eight years later in the revised tables which he appended to "The Doctrine of the Sphere". In the third book of the first edition of the *Principia* (1687), Newton did not produce a theory of the Moon's motion: he merely considered the equation of the centre (Prop. XXVIII) and the variation (Prop. XXIX), and investigated the motion of the Moon's nodes. When he began to contemplate preparing a second edition of his epoch-making work, he recognized this to be one of the main weaknesses which he required to compensate. Thus, during a visit to the Royal Observatory on 1 September 1694, he requested Flamsteed's permission to make use of the 157 lunar observations made with the latter's new mural arc between 16 November 1689 and 16 April 1692, which had been conveniently collated into three large synopses. Flamsteed agreed on condition that Newton would not impart those to anyone else, and grant him the privilege of being the first to know of any theoretical conclusions that might be deduced from them; to which Newton readily assented. Flamsteed's prime reason for insisting upon this proviso was to prevent erroneous conclusions being drawn by others from the absolute values of the Moon's right ascensions and declinations found with the aid of Tycho Brahe's ecliptic coordinates of the bright star α Arietis, which he knew to be liable to error. He himself had adopted them merely on a provisional basis until such time as he would determine the true position of the neighbouring equinoctial point from his own solar observations. Moreover, he was aware that these lunar data and the corresponding figures calculated from the above-mentioned revised Horroccian tables published in Moore's *New Systeme of Mathematicks* (1681) differed by up to $\pm\frac{1}{8}°$.

Newton wasted no time in applying these lunar observations to his theory, and a little over a month later he was able to send Flamsteed explicit instructions as to what further data he required. In the meantime, Halley had paid two visits to Greenwich. On the second occasion he was shown the same synopses and allowed to copy the maximum discrepancies. Flamsteed told him that these were partly due to the value of the eccentricity being too small which, on Horrox's principles, meant that the radius of the libratory circle (= Fb in Fig. 4) required to be made bigger in winter than in summer. Halley had, in fact, told Newton confidentially several years previously that he had devised an explanation of Horrox's seventh precept, which was similar to that of Flamsteed's given

66

above except for the fact that the libratory circle was drawn on SZ instead of on AP, and CI rather than Cx was regarded as the variable eccentricity. This modification was adopted by Newton, since it seemed to yield a better agreement with Flamsteed's observed data, and constitutes one clear distinction between their respective lunar theories. Another lay in the former's methodological approach of first discovering from his theory of gravitation what the various inequalities in the Moon's motion could be, and then determining them with the aid of the latter's data, instead of working empirically from a purely geometrical construction with no physical rationale, as Flamsteed had been doing.

What Newton still wanted were the unreduced right ascensions and meridian altitudes for each apparent time of observation with the mural arc, and Flamsteed obliged him by sending a further six lunar positions observed between 16 April and 16 June 1692, and twenty-six others pertaining to the period 15 September 1694 to 18 January 1695. In thanking Flamsteed for these last observations Newton, in a letter dated 16 February 1695, states:

> ". . . all the world knows that I make no observations myself, and therefore I must of necessity acknowledge their author: and if I do not make a handsome acknowledgement, they will reckon me an ungrateful clown. And, for my part, I am of the opinion that for your observations to come abroad thus with a theory which you ushered into the world, and which by their means has been made exact, would be much more for their advantage and your reputation, than to keep them private till you die or publish them, without such a theory to recommend them. For such theory will be a demonstration of their exactness, and make you readily acknowledged the exactest observer that has hitherto appeared in the world. But if you publish them without such a theory to recommend them, they will only be thrown into the heap of the observations of former astronomers, till somebody shall arise that, by perfecting the theory of the moon, shall discover your observations to be exacter than the rest. But when that shall be; God knows: I fear not in your lifetime, if I should die before it is done. For I find the theory so very intricate, and the theory of gravity so necessary to it, that I am satisfied it will never be perfected but by somebody who understands the theory of gravity as well, or better than I do. But whether you will let me publish them or not, may be considered hereafter. I only assure you at present, that without your consent I will neither publish them, nor communicate them to anybody whilst you live, nor after your death, without an honorable acknowledgment of their author."[5]

67

Newton duly supplied Flamsteed with a set of lunar tables just over two months later, which represented the culmination of his efforts in analysing the mural arc data. All that the latter had left to send him were his sextant observations of the Moon, and in July 1695 he communicated those made from 16 January to 28 June 1677 and offered to send the rest when Newton should have need of them. He also transmitted a catalogue of about 130 stars derived from his sextant observations rectified for the beginning of 1686, and a nonagesimary table that had been constructed by Abraham Sharp for calculating the parallactic angles necessary for the conversion of the Moon's observed places into true (geocentric) ones. Flamsteed pointed out to Newton that the right ascensions of the Moon were not observed directly in this case, but deduced from the observed differences in the times of transit of the Moon's limb and from the right ascensions of stars obtained by calculation (in accordance with the method whose principles have been outlined in Chapter 2). He calculated 30 new mural arc observations of the Moon observed during the earlier half of 1695 in order to discover how much better Newton's new tables of equations of the lunar apogee and eccentricity agreed with observations than his own tables appended to Moore's *Mathematicks*, and found that the errors were generally halved although they amounted at times to over $\pm10'$. However, this was still completely inadequate for the purposes of navigation.

Flamsteed's statement in his letter of 2 July 1695 that he had saved Newton a great deal of labour by his calculations in the lunar synopses and that it would be "but equitable and just" that his correspondent should save him some too, seems to have touched a nerve, since it brought forth the following tirade from the latter:

"Sir,

After I had helped you where you had stuck in your three great works, that of the theory of Jupiter's satellites, that of your catalogue of the fixed stars, and that of calculating the moon's places from observations; and in all these things freely communicated to you what was perfect in its kinds (so far as I could make it), and of more value than many observations, and what (in one of them) cost me above two months' hard labor, which I should never have undertaken but upon your account, and which I told you I undertook that I might have something to return you for the observations you then gave me hopes of, and yet, when I had done, saw no prospect of obtaining them, or of getting your synopses rectified, I despaired of compassing the Moon's theory, and had thoughts of giving it over as a thing impracticable, and occasionally told a friend so who then made me a visit . . ."[6]

Flamsteed described this reply in a personal memorandum as "nasty artificiall unkind arrogant".[7] The theory of Jupiter's satellites was a topic considered briefly by Newton in December of the previous year in response to a request made by Flamsteed for a quantitative estimate of the major inequality in their respective motions. The fact that Jupiter's own motion was very irregular was well known, but part of the effect was systematic. From 1664 to 1676 the mean motion found from observations agreed quite well with those calculated by William Bossley, whereas from 1676 to 1688 it appeared retarded by 12'; later, between 1688 and 1699, it was found to accelerate by 1', and from 1704 to 1710 to be retarded by 10'. Yet, when this secular acceleration was allowed for, the errors were never more than $\pm 8'$, which was twice as reliable as the predictions in Kepler's Rudolphine Tables and similar to the degree of accuracy obtained in the cases of Mars and the Moon. Newton was able by the end of 1694 to supply Flamsteed with the information that he wanted, and he assured him shortly afterwards that Saturn would have no sensible influence upon the motions of Jupiter's satellites. The help given with the star catalogue refers to his construction of the table of refraction mentioned in Chapter 2, and this was what had involved more than two months' concentrated effort. The third source of assistance had been Newton's computation of his table of the equations of the Moon's apogee and eccentricities, and that of the horizontal lunar parallaxes.

The friend to whom Newton alludes in the extract cited above may have been Richard Bentley, the classical scholar who had also been corresponding with him upon the theological implications of the new doctrine of universal gravitation, and had persuaded Newton to allow him to edit the projected second edition of the *Principia*. Bentley was apparently responsible for spreading the false rumour in London that that edition would have to be published without the lunar theory because Flamsteed had not been giving Newton the necessary data, while Halley had been hinting that the Astronomer Royal might continue to withhold these in the future. Flamsteed, still smarting from the aggressive tone of Newton's letter, justifiably complained to him about the propagation of these falsehoods. Newton quickly extracted an apology from one of those responsible and an assurance that the unjust rumour would go no further; thus restoring the amicable relationship with Flamsteed which had existed beforehand. The former's prestigious appointment as Warden of the Mint—a duty which he performed so conscientiously that it almost put an end to his scientific studies— and the latter's prolonged illness from gall-stones, caused a lull in their correspondence until a circumstance arose which imposed a further strain upon their mutual friendship.

John Wallis, having previously heard from John Caswell that Flamsteed had discovered evidence for stellar parallax, wrote to ask whether the Astronomer Royal would be prepared to draw up his observations of that phenomenon in the form of a Latin letter which he (Wallis) could insert in the third volume of his *Opera Mathematica*, subsequently published in 1699. Flamsteed readily agreed, but left that editor to make the Latin translation and supply the date (12 December 1698). In the original English version of this letter, which has unfortunately not been preserved, Flamsteed had apparently made reference to his having supplied Newton with observations of the Moon. When the latter learnt this via the intermediary of David Gregory, Edward Bernard's successor as Savilian Professor of Astronomy at Oxford, he told Gregory to ask Wallis to suppress that statement on the grounds that people might otherwise accuse him of dabbling in mathematical matters when he ought to have been devoting his full attention to his new duties as Master of the Mint, and that he might be unable to perfect and publish the lunar theory in any case. Wallis told Flamsteed of Gregory's request, but without mentioning that it originated from Newton himself; with the result that Flamsteed wrote twice to Newton informing him of these circumstances before receiving a rather cool reply reprimanding him for daring to mention names in print without first requesting permission to do so.

Flamsteed answered that the observations had been made at the King's own Observatory; Halley had already been telling people that Newton's theory was complete, and everyone knew that Newton himself made no astronomical observations. It had been Newton's request for the latitude of the Greenwich Observatory which had led Flamsteed, early in 1695, to collect his observed meridian zenith distances of the Pole star into a synopsis, and thereby to detect the systematic tendency for these distances to be greater between the summer solstice and autumnal equinox than just before and around the time of the winter solstice, an effect which he interpreted as evidence for a stellar parallax of about $\frac{1}{2}'$ in accordance with the view expressed in his letter to Seth Ward fifteen years earlier (cf. Chapter 2). Thus he no longer regarded Sirius as a suitable reference point for the determination of sidereal time, and did not concede to Newton's suggestion that transit observations of that bright star should be used in preference to measurements with his mean-time clock for the timing of celestial events.

The publication of Flamsteed's letter elicited a quick response from Jacques Cassini in Ulricus Junius's *Novae et accurate motuum coelestium ephemerides* (Parisiis, 1700), reprinted in Bernard de Fontenelle's *Histoire de L'Académie des Sciences*. Cassini first summarizes the main features of Flamsteed's data and then explains why these are actually incompatible with the hypothesis of a stellar parallax. Gregory, on the other hand, in his *Elementa Astronomiae* (Oxoniae, 1702), does not take issue either with the data or with the inferences drawn by Flamsteed from them, but makes the alternative proposal that the Pole star is nearer to the north celestial pole around the time of the winter solstice since the Earth's southern hemisphere is denser than its northern one. As Flamsteed points out in a letter to Caswell on 5 September 1702,[8] however, such a conjecture is opposed to Newton's mechanics, the land mass in the northern hemisphere being in excess of that in the southern, earth and rocks denser than sea-water, and hills farther from the Earth's centre than the surface of the ocean. Even if one were to suppose that the southern hemisphere were heavier, no new nutational effect would be implied other than what would otherwise have existed, there being merely a translation in the position of the centre of gravity. He thought that Gregory had been misled by the assertion by Christian Huygens in *Systema Saturnium* (1659) that the diameters of the fixed stars are insensible—an erroneous view, in his opinion, resulting from the fact that the objective of the Dutchman's telescope was lightly covered with smoke to reduce the glare of the Sun, the particles of which intercepted the light rays from brighter stars and caused them to appear smaller and dimmer. Newton's demonstration thirteen years later that not all rays could be brought to a single focus after being transmitted through a refracting medium had meant that even if heavenly bodies were actually pinpoints of light they would still appear to have small but measurable angular diameters. Theoretically, this 'circle of least confusion' (as it became called) would be 5″ to 6″ for a telescope of 2-inch aperture and 20 to 30-foot focal length. Yet, in practice, Flamsteed found that at least two-thirds of the observable stars were less than 1″ in diameter whereas others appeared to be between 6″ and 12″; thereby implying that some stars did have measurable diameters. Gregory's alternative suggestion that the annual stellar parallax could be found from observations of the angular distance between two stars at different times of the year was impracticable because of the greater inaccuracy of measuring a difference in the meridian altitude of two stars with a movable instrument than in that of one star with a fixed instrument.

The appearance in Gregory's book of Newton's lunar theory, described in the latter's own words, revealed at once to Flamsteed that Newton had not respected his wish that he should be the first to learn of any theoretical results obtained from the application of the data that he had communicated. This confirmed earlier rumours spread by both Gregory and Halley, that Mr Newton had perfected the theory of the Moon. In addition to the annual equation, equation of the centre (including the evection), and the variation, Newton introduced four new equations but failed to demonstrate how he had obtained them. William Whiston, in his *Praelectiones Astronomicae* (1707), suspected that one, at least, had been inferred from Flamsteed's observations (from January to July 1677) and preferred to print the latter's revised version of Horrox's lunar tables *without* incorporating anything from Newton's theory. Flamsteed himself was well aware that Newton's 'new' theory of the Moon was little more than an amended version of Horrox's, but was not convinced that it was much of an improvement since the sixth and seventh equations did not agree with observations. After seeing a modified version of Newton's rules in the Scholium following Prop. XXXV in Book 3 of the second edition of the *Principia* (1713), he realized that one reason for this disagreement had been a mistake initially incurred by Newton in the sign of the sixth equation. In the third edition of 1728 the seventh equation was omitted altogether and a new formula proposed for the correction of the variation.

As far as the practical astronomer was concerned there was little to recommend any of these versions of Newton's lunar theory, since Flamsteed's own lunar tables of 1681 and those in Halley's edition of Thomas Streete's *Astronomia Carolina* (1710) yielded predictions of a similar degree of reliability; namely, to around $\pm 8'$. In the appendix to this book, Halley presents a proposal for finding longitude at sea by the method of lunar distances—a subject which had occupied his attention as early as 1682 when he embarked upon a series of observations of the Moon and planets with a 5½-foot radius sextant at his home observatory at Islington. His earlier experience had convinced him that this was potentially the method best suited to navigation at sea. Since he was well aware that an error in the lunar tables would be repeated at the corresponding phase of any succeeding Saros cycle of 223 lunations (= 18 years 11 days), he accordingly resolved to make "a sedulous and continued Series of Observations to be collated with the Calculus, and the Errors noted in an Abacus"[9] from which at all times in a similar situation of the Sun and Moon the appropriate correction could be extracted. He naturally used at that time the lunar tables of Horrox revised by Flamsteed, that had just been published one year earlier. Within sixteen months he had observed on 200 days, but the unexpected suicidal death of his father and the resulting shock to his own fortunes forced him to interrupt this plan.

Perhaps this was partially responsible for Halley then transferring his attention to the terrestrial method of longitude determination depending upon the variation of the magnetic needle. On his voyage to St Helena in 1676 to observe stars in southern constellations, the results of which were duly embodied in his *Catalogus Stellarum Australium sive supplementum Catologi Tychonici* (1679), he had observed that the magnetic dip vanished when he was 15° north of the Equator. This discovery had encouraged him to prepare a synopsis of well-established measurements of magnetic variation at places of known latitudes and longitudes, which revealed the total inadequacy of previous theories to explain this phenomenon, proposed by William Gilbert and René Descartes; and even of Henry Bond's attempt to account for the secular changes in both the magnetic variation and dip by assuming that two diametrically opposed terrestrial magnetic poles lagged behind the Earth in its daily rotation so as to describe circles about the geographical poles, to which Robert Hooke had also lent his support. Halley felt that a better agreement with these magnetic measurements would be obtained by supposing the Earth to have two magnetic poles near each geographical pole, and expounded this view in the *Philosophical Transactions* for 1683 when he proposed approximate locations for these four magnetic poles. He suggested how the isogonics would be arranged in the North Pacific area but stressed the need for more such measurements being made on land to ascertain how the magnetic force varied with distance from the attracting pole. The existence of the secular change indicated, on this hypothesis, that one pair of poles moved relatively to the other pair.

Up to this time, Halley had managed to stay on civil terms with Flamsteed, although there are hints in Hooke's *Diary* for 7 January 1676 and in a letter from Flamsteed to Bernard on 8 February 1678[10] that the relationship was based more on mutual respect for one another's abilities rather than on any feelings of friendship. Temperamentally, the two men were totally different. Halley was healthy, an extrovert, adventurer, jocular in conversation, and atheistically inclined. Flamsteed was none of these things, and disapproved of Halley's close association with Hooke, for whom he had a strong personal dislike. However, these are reasons for their lack of intimacy, and scarcely a basis for the bitter animosity which grew up between them and which was particularly strong on Flamsteed's side. The above-mentioned theory of terrestrial magnetism would appear to have provided the spark which kindled the feeling of hatred that was destined to produce an irreconcilable rift in their relationship. After reading Halley's paper on magnetic variation, Flamsteed openly and unjustly accused him of plagiarizing the views of his late friend Peter Perkins from unpublished manuscripts purchased for a small fee from the latter's widow within a fortnight of his decease on 12 December 1680. As Flamsteed tells Towneley on 4 November 1686:

"His [Halley's] discourses ... concerning the variation of the needle and the 4 poles it respects I am more than suspicious was got from Mr Perkins, the master of the mathematical school at Christ Hospital, who was very busy upon it when he died and often told me that the needle did not point to any particular and certain pole but to respective points (which we find betwixt Mr Halley's poles) on our Earth. Mr Halley was frequently with him and had wrought himself into an intimacy with Mr Perkins before his death and never discoursed anything of his 4 poles till some time later I found it published in the transactions. I charged him modestly with it, but he made no answer to the purpose, knowing that Mr Perkins had a paper entered on the Journal of our Society, whereby he may be convicted of his cunning. ... He delays his publishing the discourse he read to our Society concerning the monsoons or trade winds, pretending that he stays for the arrival of an ingenious seaman who

74

is very well acquainted with them and daily expected him, but a Dr of physic says he has only translated a treatise of Monsieur Bardelot on that subject with some variations and suspects the discovery thereof prevents him. He is got into Mr Hooke's acquaintance, been his intimate long, and from him he has learned those and some other disingenuous tricks for which I am not a little concerned for he has certainly a clear head, is a good geometrician and if he did but love labor as well as he touts applause, if he were but as ingenious as he is skillful, no man could think any praises too great for him. I used him for some years as my friend and I make no stranger of him still, but I know not how to excuse these faults, even in my friends. Since he ran into Mr Hooke's designs and society, I have forborn all intimacy with him."[11]

This quotation sums up the situation at that time, but within three weeks of the letter being written Flamsteed had a fresh cause for complaint which he recounts in four letters to William Molyneux of Dublin during the winter of 1686–7.[12] According to this evidence, it would appear that a certain Mr Dee with whom Flamsteed professed to be unacquainted, had been criticizing Flamsteed in some coffee-house for being a mean host. This charge probably originated from Flamsteed's deliberate inhospitality to Halley, Thomas Molyneux (his correspondent's brother), and Charles Boucher, during a visit by that trio to the Royal Observatory. The intention had been to repay Halley for the casual manner in which he had treated Flamsteed on three occasions when the latter had stayed at Halley's London home, despite the Astronomer Royal having entertained his younger associate on numerous occasions at Greenwich. Boucher had even gone so far as to ask Flamsteed why he did not resign his office in order to let Halley take over. Even though he suspected that Halley himself had instigated these jibes out of envy for his position, Flamsteed had nevertheless given him a catalogue of all his observations on the eclipses of Jupiter's satellites for publication in the *Philosophical Transactions*. However, he strongly disapproved of an editorial comment by Halley when publishing his tables of the high tides at London Bridge in the 1686 issue of that periodical, to the effect that Flamsteed's generalization of these results to other sea-ports (by allowing for a mere additive or sub-tractive correction) was not justified. The grounds for this remark was the disagreement between Flamsteed's prediction for Dublin and the tides observed there, information regarding which Halley obtained from William Molyneux. In complaining to the latter, Flamsteed writes, regarding Halley:

> "I was never so unworthily dealt with by any before, & I thought hee was obliged to be more civil to me than to any, but I see obligations are thrown away upon some p[er]sons. I am sensible hee designes if hee can continue the Injury & am therefore resolved not to commit any thing againe into his hands."[13]

He warns his correspondent to send his own contributions to the *Philosophical Transactions* in future to the President of the Royal Society or to arrange for them to be read by his brother at one of the weekly meetings at Gresham College, in order to avoid the risk of editorial misrepresentations. He threatened to make Halley's malice and errors public unless the latter placed more faith in the experience of all knowledgeable seamen and retracted his criticism of the tide-tables; and when Halley refused to do so, he angrily reproached him with his former faults. Another thing which seems to have unduly disturbed him was that the account of the tides with Halley's reflections upon it appeared at the front of the volume for 1686, whereas Flamsteed's catalogue of eclipses happened to be situated near the middle.

For such reasons, and because of "some very foolish behaviour of his [Halley's] very lately", Flamsteed told Newton on 24 February 1692 that he had "no esteem for a man who has lost his reputation both for skill candor & Ingenuity by silly tricks ingratitude & foolish prate".[14] He felt under no obligation to be governed by Halley's and his friends' wishes that he would quickly publish his Greenwich observations, and was quite capable of organizing his own affairs. Not long afterwards, after Halley had given a more detailed analysis of his theory of magnetic variation postulating a differential rotation of the Earth's core relative to a surrounding shell as the reason for the observed secular change, Flamsteed renewed his charges of plagiarism and continued to criticize him to his face and in letters to Molyneux, Newton, Sharp, and others. Halley, for his part, was annoyed at Flamsteed for publishing as his own certain lunar eclipse observations which he had made and calculated, and becoming exasperated at the repeated rebuttal of his own efforts to effect a reconciliation. As Flamsteed tells Caswell on 25 March 1703, referring to Halley:

> "last time I saw him many words passed betwixt us; he complained of my unkindness highly, and asked loudly what he must do to gain

my friendship; I answered roundly *he must become a just, serious, and virtuous man*, and then I should be his friend immediately. The answer the company took notice of, but he passed [it] over. I am of the same mind still, and if he can be such, there is no need of any reconciliation; if he cannot, it would be highly to my prejudice to make an accommodation with one whom I never willingly injured."[15]

The advice would appear not to have been taken, however, since Flamsteed tells Sharp at the end of that year that Halley "now talks, swears, and drinks brandy like a sea-captain".[16]

This comparison was not inappropriate, for Halley had actually been a sea-captain from 1698 to 1700 in charge of a ship called the "Paramour". His mission had been to observe the variation of the magnetic compass in the South Atlantic, to determine the geographical coordinates of all ports and islands visited, to seek new lands, and to visit the English West Indian plantations on his return voyage. As a result of having had a rebellious first lieutenant, Halley had been obliged to cut short his first voyage and return to England to obtain a new crew, and to set out for a second time in September 1699. In his journal of this voyage, Halley recorded the estimated positions of his ship from day to day; used eclipses of Jupiter's satellites and appulses of the Moon to stars for determining longitudes at sea; recorded temperatures, pressures, and the force and direction of the wind. He took one or more magnetic observations almost every day, making appropriate allowances where necessary for atmospheric refraction and the dip of the horizon. A modern re-analysis of his results has confirmed that the variations were reliable to well within 1°. The principal products of the voyage were two isogonic charts (for the epoch 1700) of the Atlantic area and of all the oceans of the world. Revised editions of Halley's World Chart of 1702 taking account of the slow changes in magnetic variation were issued throughout the eighteenth century but found to be of less value in navigation than their author had hoped.

William Whiston criticized Halley for not having made measurements of the magnetic dip at the same time. James Pound and James Cunningham had, however, done so on a voyage to China in 1700; the latter had published their results in the *Philosophical Transactions*. Unfortunately, these data did not agree with others obtained later. The deficiency was compensated by Père Noel's accurate observations on his voyage from Lisbon to the East Indies in 1706, and those of Feuillée made near the South American coast to Lima (Peru) in

77

1708. It was his study in 1718 of Eberhard's supposed method of finding longitude by the dipping-needle which caused Whiston to give serious consideration to this matter and to begin experimenting with such an instrument. Whiston, towards the end of his Historical Preface to *The Longitude and Latitude found by the Inclinatory or Dipping Needle* (London, 1721), completed on 20 February 1719–20, expresses the hope that he will not be denied the rewards which nations had set aside for the discovery of longitude at sea for the method which he is proposing, and alludes to "a like Occasion formerly" when he and his now deceased friend Humphrey Ditton had entertained the same aspiration.

The occasion in question had been their unsuccessful bid to interest a House of Commons Committee set up in 1713 as a result of a petition by sea-captains and London merchants to provide a public reward for the discovery of longitude at sea, in a scheme to establish floating lightships at fixed points on principal maritime trading routes, from which rockets were to be fired at regular intervals and arranged to explode at a specified height of 6440 feet above sea-level. The distance between the two ships was to be estimated from the observed time-lag between the flash and the sound of the explosion. The proposal which they published in 1714 under the title "New Method of Discovering the Longitude", was referred by the Committee to the expert judgment of Newton, Halley, Samuel Clarke, and Roger Cotes, by whom it was deemed to be impracticable and useful only inasmuch as it helped a mariner to keep account of his longitude rather than to discover it (if lost). Newton also informed the Committee that the known methods of longitude determination involving Jupiter's satellites, the Moon's meridian transit or distances from bright zodiacal stars, or the use of a reliable timekeeper, were also quite impracticable for the purpose of accurate oceanic navigation; whereupon it was resolved

> "That a Reward be settled by Parliament upon such Person or Persons as shall discover a more certain and practicable Method of ascertaining the Longitude, than any yet in Practice; and the said Reward be proportioned to the Degree of Exactness to which the said Method shall reach."[17]

A Bill was duly drafted between 11 and 16 June 1714 by nine members of the Committee and presented to the House on three occasions before being passed with all the approved amendments by the House of Lords less than a month later. Under the terms of this Act (12 Queen Anne, cap. xv) the respective awards were to be

£10,000 if the method were accurate to within 1°, or 60 nautical miles

£15,000 if the method were accurate to within $\frac{2}{3}$°, or 40 nautical miles

£20,000 if the method were accurate to within $\frac{1}{2}$°, or 30 nautical miles

The magnitudes of these sums reflect both the urgency and the intrinsic difficulty of achieving a satisfactory solution. A body of (initially) 22 Commissioners, known collectively as the Board of Longitude, was appointed to authorize the payment of these sums: it included high-ranking Admiralty officials, politicians, and scholars, all of whom had a concern in the safety of H.M. ships and crews and the improvement of Britain's maritime trade. The Board was empowered to pay one-half of the appropriate bounty to whoever could invent a method which a majority deemed to be both 'useful and practicable'—a somewhat ill-defined phrase which was destined to give rise to some heated controversy later in the century. The other half was to be paid as soon as a ship using that method should sail from a port in the British Isles to another in the West Indies without erring in longitude by more than the specified amounts. Smaller payments, not exceeding £2000, could be paid to assist experiments likely to yield useful results. Financial and administrative matters were to be referred to the Admiralty Office and warrants paid through the Treasurer of the Navy, the chairman of the Board being the First Lord of the Admiralty. Legally, however, the Board was responsible solely to Parliament. Among the first fortune-seekers was the Welshman Zachariah Williams, who placed his faith in the predictive reliability of the phenomenon of secular change in magnetic variation. In fact, he never formulated the tables which he promised the Royal Society he could make if given suitable financial encouragement. Some consideration was also given to Henry de Saumarez's so-called 'Marine surveyor' or mechanical ship's log; but this was incapable of correcting errors due to oceanic currents and, though ingenious, of little value. A few 'cranks' also submitted absurd proposals. It was not until 1737 that the Board met in order to consider the serious claim of John Harrison for financial encouragement to enable him to continue work on his first marine timekeeper.

Halley, as Savilian Professor of Geometry at Oxford, was a member of the Board at the time of its creation; and, as a well-known expert on astronomical and nautical matters, undoubtedly the member best qualified to pass judgment upon the feasibility of the initial proposals which came to it. However, he himself had hopes of qualifying for the tempting award and was accused of dealing somewhat too harshly with the sincere, albeit misguided, proposals of others. His editing of the third edition of Thomas Streete's *Astronomia Carolina* (Londini, 1716) provided an opportunity for reiterating the wish expressed six years previously in the second edition that his proposal for keeping account of the errors in the lunar theory might "find a kind Acceptance from those that it chiefly concerns".[18] It must be stressed that this statement did not apply initially to the Board of Longitude which came into existence only in 1714, and should therefore not be construed as an attempt by Halley to win one of the longitude prizes.

The death of Flamsteed and Halley's appointment as successor on 9 February 1720 placed the latter in a position to begin this abandoned line of research anew. He was well aware that Newton's lunar theory had been founded mainly on observations made when the Moon was in conjunction with, or in opposition to the Sun; and that it was of particular importance to establish the errors in the quadrantal and octantal phases of its orbit, thereby facilitating "the too laborious Calculations of the Moon's place *extra Syzygias*" (as he puts it in his appendix to Streete's book). But he had other problems with which to contend before he could turn his attention to beginning an 18-year lunar ephemeris—an ambitious task to undertake for a man already in his sixty-fourth year!

According to Joseph Crosthwait, Halley came to Greenwich early in February 1720, and gave Mrs Flamsteed only three days in which to remove herself and her belongings from the Royal Observatory. He must have been shocked to find, on his next visit, that this left him with no major instrument and no furniture! This action was legally justifiable, since the instruments had either been presents to his predecessor from Sir Jonas Moore or made, erected, and repaired at Flamsteed's own expense. Halley was fortunate in obtaining £500 to re-equip the Observatory, which enabled him to purchase from George Graham a new transit telescope and 8-foot radius mural quadrant; however, the time required for constructing and mounting these instruments meant that it was October 1721 before his first recorded observation could be made. He was, of course, seldom idle, and no fewer than twelve scientific papers by him were published during this and the previous year—between his resignation as editor of the *Philosophical Transactions*, and that of his office as secretary to the Royal Society. These included a firm rejection of Jacques Cassini's claim in the Paris Academy's *Mémoires* for 1717 to have detected the annual parallax of Sirius, which Halley attributed to the variable effect of refraction. No doubt it was his examination of that question which induced him to publish "Some Remarks on the Allowances to be made in Astronomical Observations for the Refraction of the Air" only a few months later, along with an extract from the accurate table of refractions which Newton had prepared for Flamsteed in 1695, for altitudes between 0° and 75°.[19] Another study of direct relevance to his work at Greenwich, made by Halley during the winter of 1720–21, was an analysis of the errors incurred in the use of a transit instrument with a micrometer eyepiece in conjunction with a pendulum clock, to find differences in right ascension and declination. He adopted Jacques Cassini's arrangement of having four hairs intersecting at angles of 45°, their common centre being that of the eyepiece. The instrument was adjusted so that one star was carried by the diurnal rotation through that centre along one of these wires. The difference in right ascension was then deduced from the interval of time taken by a following star to cross the hair at right angles to that one; and the difference in declination from that taken by it to pass from one diagonal thread to the others. This method was useful not only for stellar observations but for determining the positions of the Moon and planets by measuring their appulses to the stars. In a brief article on this latter subject written later that same year, Halley asserts that no other celestial observation was "so capable of perfect exactness".[20]

Baily, in his account of Halley's astronomical observations at Greenwich, criticizes him and his contemporaries for paying too much attention to such observations and too little to improving star positions—an attitude which he claimed was reflected in

"the conduct of Newton and Halley in having directed that the observations of Flamsteed should be so published as to include only those parts relating to the moon and planets; and leaving the mass of his observations of the fixed stars wholly unnoticed".[21]

But this was primarily a matter of expediency dictated by the necessity of producing a more reliable basis for accurate oceanic navigation—the urgency of which had just been high-lighted in the public eye by the wreck of Vice Admiral Sir Cloudesley Shovell's fleet off the Isles of Scilly (1707) with the consequent loss of some 2000 lives. Halley's decision to spend less time making stellar observations at the Royal Observatory was undoubtedly conditioned by the fact that he had edited Flamsteed's star catalogue (1712) and that Crosthwait and Sharp would soon be producing the other (which duly appeared in 1725). Thus he informed the Royal Society just over three months after his appointment that all he intended to do in this area of study was

> "to enlarge the British Catalogue of fixed stars, by inserting in the vacant spaces of the zodiac all such stars as are plainly visible through a telescope of five feet length; in order to make the method of finding the longitude at sea by the moon more practicable than it is at present."[22]

Halley began observing towards the end of 1721, and continued until shortly before his death on 14 January 1742. Four of his five quarto vellum-bound volumes of observations, covering the period between 1 October 1721 and 31 December 1739, are still extant, but 'a *fifth* volume of observations' presumably containing the continuation of these data between 1 January 1740 and January 1742 was not found by Nevil Maskelyne when he later came to compose a memorandum concerning Halley's manuscripts. The extant volumes were in such a state of confusion, and contained so many computations and extraneous matter, that Baily caused a 518-page transcript to be prepared, which was presented by the Lords Commissioners of the Admiralty to the Astronomical Society of London on 14 December 1832 and is still preserved in the archives at Burlington House, London. These observations have never been published; only the *Tabulae Astronomicae* derived from them were published posthumously in London in 1749. Accounts of particular phenomena such as a solar eclipse on 27 November 1722, a transit of Mercury on 29 October 1723, lunar eclipses on 18 June 1722 and 15 March 1735/6, lunar and magnetic variation methods of finding position at sea, appeared in the *Philosophical Transactions* for the years 1722, 1723, 1725, 1732 and 1737. The transit of Mercury, predicted by Halley in 1691 and now observed by him, when compared with that which he had observed in 1677, served to establish an improved value for the planet's mean rate of motion and to fix the position of the node where its orbit crosses the ecliptic.

The minute-books of the Council of the Royal Society for 12 May 1726 reveal the state of the Royal Observatory at that time, only several months after the mural quadrant had been erected:

"As to the state of the Observatory.

That all the instruments now lodged or erected at the Royal Observatory, and belonging to it were procured by the present Professor [Halley]: those which were used by his predecessor being carried off and claimed by his executors.

That there is a room adjoining to the west end of the house, newly erected, which serves as an observatory for taking the transits of objects on the meridian. It being furnished with a curious telescopic instrument of five-feet radius, fitted to an axis, and adjusted with screws to revolve in the plane of the meridian. And a plain week-clock, standing by it, for making the observations.

That the great room in the observatory is furnished with a plain month-clock, and three very good telescopes: one of nine feet, another of sixteen, and the third of twenty-four feet in length. And also two very good micrometers of different forms.

That in the garden, from off the south-east corner of the house, there is erected another building, being a room designed to hold two large mural quadrants of eight-feet radius, for observing the meridional altitudes of objects. One of which quadrants is to command that part of the meridian which lies to the south: and the other, that part which lies to the north. And in the middle of the said room is erected a firm stone-wall, lying north and south, being eleven feet high, nine feet long, and two feet thick; consisting of nine large stones cemented. To the east face of which wall is affixed the large quadrant, which is for taking the observations to the south; being entirely finished and fitted up for use.

That they are informed that the materials for making the other quadrant are procured, and many of its parts formed: at the brass limb, the iron bars, and the tube for telescopes, with some others."[23]

After the Board of Visitors (consisting of Dr Taylor, Dr Stukeley, John Hadley, George Graham, Martin Foulkes) had inspected the new instruments and the accounts four days later, they reported that a substantial instrument for observing *outside* the meridian was still required and that about £200 more than the original £500 allocation would be necessary for its purchase; however, this was not to be.

The most serious sources of error affecting the absolute accuracy of all Halley's data were the poor quality of his various clocks; his failure to distinguish clearly in his observation books which clock (hence which instrument) had been used in each observation; and his failure to record whether he used all three parallel wires or only the centre one on his transit instrument when timing that event. He would appear to have possessed at least four different clocks: the 'horologium majus', used in the Great Room in conjunction with a portable telescope, the 'horologium minus', used in the Transit Room in conjunction with the transit instrument, the 'horologium murale', used in conjunction with the large mural quadrant after the installation of that instrument on 16 December 1725, and a 'monitor' journeyman or alarm-clock. None of these clocks was compensated for the effects of temperature, nor provided with a fusée to prevent it stopping when being wound up. Thus they stopped frequently—on one occasion, even during an observation— and were very unreliable. One instance recorded by Halley on 23 May 1730 was that he was obliged to put back a clock by one minute within ten days, and to shorten its pendulum to beat two seconds less per day. Halley's initial corrections for clock-errors in telescopic observations made outside the meridian were deduced by linear interpolation between the errors at local apparent noon on two successive days. However, after the installation of the mural quadrant, he adjusted the 'horologium murale' to beat sidereal time: and made a similar modification to the 'horologium minus' on 1 October 1726.

Halley's observations of both transits and zenith distances with this quadrant showed that its plane was always within 7″ of the meridian. Near each equinox between the Spring Equinox of 1726 and the Autumnal Equinox of 1728 inclusive, he made 96 solar observations with both the quadrant and the transit instrument in order to establish the zero point of the right ascension scale, and in August 1726 he made 25 observations of bright stars such as Aldebaran, α Orionis, Procyon, Sirius, and Arcturus at different altitudes. The Pole star was seldom observed. Although he made numerous observations of the superior planets, several of Venus and Mercury, and noted the circumstances of five solar and four lunar eclipses, not once did he record an eclipse of Jupiter's satellites. His energies were directed primarily towards his meridian transit observations of the Moon, of which, by the time that he composed "A Proposal of a Method for finding the Longitude at Sea within a Degree, or twenty Leagues" at the end of 1731 (i.e. one whole period of the Moon's apogee later), he had collected almost 1500—about four for each degree of the Zodiac. This proposal is a repetition of that published in the appendix to the second and third editions of Streete's *Astronomia Carolina*. He claims optimistically that the differences between the predictions of the revised lunar theory published in the second edition of Newton's *Principia* (1713) and Flamsteed's lunar observations seldom exceeded $\pm2'$, and blames the latter for incurring errors which were a "good Part of that Difference" since these were found to be at least $\pm5'$ at the third and fourth quarters of the Moon's orbit where his predecessor had made virtually no observations. Halley could not, however, resist the temptation of hitting out at his deceased antagonist:

> "Mr Flamsteed was long enough possessed of the Royal Observatory to have had a continued Series of Observations for more than two Periods of eighteen Years; by which he had it in his Power to have done all that could be expected from Observation, towards discovering the Law of the Lunar Motions. But he contented himself with sparse Observations, leaving wide Gaps between, so as to omit frequently whole Months together; and in one Case the whole Year 1716. So that notwithstanding what he has left us must be acknowledged more than equal to all that was done before him, both as to the Number and Accuracy of his Accounts; yet for want of an uninterrupted Succession of them, they are not capable of discovering, in the several Situations of the Lunar Orbit, what Corrections are necessary to be allowed, to supply the Deficiencies of our Computus."[24]

We have already described the numerous difficulties with which Flamsteed had been obliged to contend during his years at Greenwich, and their detrimental effects upon his observing programme. He had his duties as perpetual rector of Burstow to attend to, which took him away from his scientific pursuits for several weeks each year, and he suffered from frequent painful attacks of gallstone. In 1716 he was very busily engaged on the stellar observations and laborious calculations that were necessary in order to bring his catalogue to completion which, by itself, is sufficient excuse for his neglect of the lunar observations. Nevertheless, Halley still had over 800 of these to supplement the 1500 which he himself collected, which sufficed to verify that his earlier proposal of a recurrence of a similar pattern of errors in each successive Saros cycle was correct. Further confirmation was also to be found in the agreement to within $\pm1'$ between the errors found in 1721 and 1722 with his

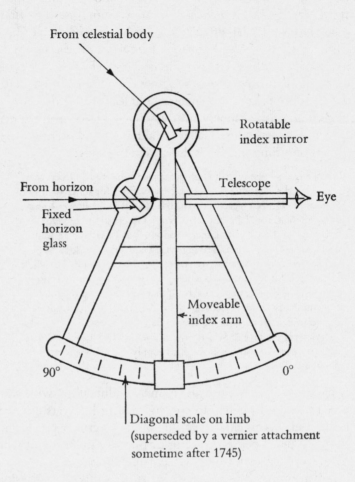

From celestial body

Rotatable
index mirror

From horizon

Telescope

Eye

Fixed
horizon
glass

Moveable
index arm

90°

0°

Diagonal scale on limb
(superseded by a vernier attachment
sometime after 1745)

Fig. 5

most recent data of 1730 and 1731. Halley's slightly more cautious conclusion was that he could in future, by the means outlined in this proposal, find the Moon's celestial position to within $\pm 2'$, and hence the longitude at sea to within 20 leagues at the Equator or 15 leagues in the British Channel. John Hadley's recent invention of a doubly-reflecting octant promised to fulfil the condition that the measurements themselves could be made with the requisite accuracy.

The great advantage of Hadley's design, over earlier attempts by Robert Hooke, Thomas Streete, and Isaac Newton to design a reflecting instrument of this nature, was that it enabled the image of a heavenly body to be seen in the eyepiece of the viewing telescope simultaneously with the horizon. Unlike the Davis quadrant, then in widespread use, it was capable of adjustment so that instrumental errors could be removed. Its principles can be easily understood with the aid of Fig. 5. The index mirror (or speculum) is rigidly attached to the index arm, and moves with it about the same axis. In theory it should be parallel to the fixed horizon glass when the pointer on the end of that arm is at the zero of the ivory or brass scale embedded into the limb of the instrument; but, if not, the appropriate instrumental correction could easily be applied. The upper half of the horizon glass is clear and the lower half covered with a metal speculum that reflects the light already reflected off the index glass, through the telescope and into the eye of the observer. Because of this double reflection, the angle measured on the scale is only half of that between the two celestial bodies being observed; thus, in order that celestial altitudes and lunar distances could be read off directly, the 45° arch was graduated from 0° to 90°. Hence many English astronomers and navigators preferred to speak of the Hadley *quadrant*, although the word *octant* seems to have been generally used on the continent of Europe. In order to observe the altitude of a celestial body, the mariner had to hold the instrument in a vertical plane and move the index arm forwards (i.e. away from him) until it attained the position where the doubly-reflected rays of light from the body and those received directly from the horizon were seen to coincide. The reading on the scale was then the required altitude. The angular distance between any two celestial objects (e.g. the Moon and bright zodiacal star) was found in a similar manner by tilting the instrument into the plane defined by these two bodies and the observer's eye.

In August 1732, John Hadley wrote to the Admiralty describing his octant, with the result that instructions were issued that it should be tested on board a 60-ton yacht called the 'Chatham'. There was, of course, no question of his applying for one of the longitude prizes, since his invention had no bearing on the fundamental problem of finding *standard* time even though it ensured that a ship's latitude could be reckoned with a much higher degree of accuracy than had hitherto been the case. The observers found Hadley's own wooden model more convenient to handle than a brass model constructed by the London mathematical instrument-maker, Jonathan Sisson; despite bad weather and the novelty of the instrument, they considered it to be capable of measuring altitudes and lunar distances with a consistency of within 1′. Significant systematic errors were nevertheless often incurred in practice because of the mariner's failure to apply corrections for astronomical refraction and the dip of the horizon: these conspired to produce observed values of the noon altitudes which were too high and the principal cause of the inconsistent values for the Sun's declination published in seamen's almanacs. Yet another source of uncertainty was involved in applying the Sun's tabulated declination to the observed meridian altitude in order to obtain the latitude, for in general no correction was made to compensate for the hourly change in declination during the time-interval corresponding to the ship's longitude difference east or west of the meridian for which the tables had been prepared.

Hadley's instrument quickly appeared on the commercial market, and was described in a pamphlet published in 1734. The diagonal method of graduation was used to obtain a precise reading from the arc, but at some time soon after the expiry of Hadley's patent eleven years later, a vernier scale began to be substituted in its stead. A second horizon glass, fitted at right angles to the first, enabled 'back observations' of angles greater than 90° to be made by using the horizon in the opposite direction from the Sun. Later modifications included: telescopic sights, non-tarnishable glass or silvered reflecting surfaces, a filter to reduce glare from the Sun, a tangent screw for fine adjustment, a magnifying lens, a rotatable index or horizon glass, and sometimes even an artificial horizon so that observations could be taken in misty weather.

Seven years after Halley's death, his *Tabulae Astronomicae* was published in London by John Bevis. These tables, with precepts in both English and Latin, had been submitted to the press by their author as early as 1717 and printed off two years later—before he became Astronomer Royal. In fact, it had been as a result of this appointment that Halley decided to defer their publication so that the lunar tables could be compared with the results of his intended Greenwich observations and incorporate the necessary corrections. One feature which Halley introduced into the longitude of each planet was a 'secular equation' (i.e. a term dependent upon the square of the time measured from a selected epoch) to take account of the progressive and unexplained changes in the mean motions of Jupiter and Saturn—a phenomenon which he correctly ascribed to the effects of mutual gravitational attraction in disturbing the centri-petal forces of the Sun, to which we shall be referring again later. On the other hand, the tables, having been prepared prior to the important discoveries of aberration and nutation which we shall be discussing in the next chapter (Chapter 5), made no quantitative allowances for the influences of these two phenomena on the apparent positions of heavenly bodies. Similarly, no corrections were made to zenith distances (hence, to declinations) to compensate for the fact that the Earth's shape approximates to a spheroid, in accordance with Newton's prediction in the *Principia*, since this was firmly established only in 1742 after the return of French scientific expedi-tions to Quito in Peru (the present-day Ecquador) and Lapland. For such reasons, these tables made little impact on the progress of positional astronomy.

NOTES

1. Flamsteed's transcription of nine letters from Crabtree to Gascoigne, and ten letters from Gascoigne to Crabtree, was begun on 25 January 1671/2 and completed on 18 May 1672. It is preserved among his manuscripts at the R.G.O., P.R.O. Ref. 40, ff. 9^v–22^v.
2. *Ibid.*, f. 9^v.
3. *Ibid.*, f. 19^r.
4. This precept, which was taken by Horrox from Kepler, is stated in *ibid.*, f. 18^v as follows:—"from y^e ☽^{s} place first æquated take y^e suns place, by that elongation find Tychoes variation (Tab Rudolphi. pag 82) of which take $\frac{9}{10}$, & adde it in y^e ☽^{s} place before equated, so have y^u y^e moones true place in her orbe which reduce to y^e ecliptick."
5. Scott (Ed.), pp. 87–88.
6. Newton to Flamsteed, 9 July 1695, *ibid.*, p. 143.
7. *Ibid.*, p. 144.
8. Baily, 1835 a, pp. 205–208.
9. Streete, 1716, Appendix, p. 68.

10. Baily, 1835 a, pp. 667–668.
11. Quoted from Eggen, pp. 213–214.
12. These form part of a collection of seventy letters between Flamsteed and Molyneux preserved in the Southampton Civic Record Office (Ref. D/M. 1/1), which will shortly be published by Mansell Ltd., in the first of three volumes of Flamsteed's Correspondence.
13. Flamsteed to Molyneux, 17 January 1687, quoted in Eggen, p. 217.
14. Baily, 1835 a, p. 133.
15. *Ibid.*, p. 213.
16. *Ibid.*, p. 215.
17. *Journal of the House of Commons*, **17** (Friday, 11 June 1714), 678.
18. Streete, 1716, Appendix, p. 70.
19. *Phil. Trans. R. Soc.*, **31** (1720–1), 169–172.
20. *Ibid.*, p. 209.
21. Baily, 1835 b, p. 181.
22. *Ibid.*, p. 171.
23. *Ibid.*, p. 173.
24. Halley, 1731–2, p. 188.

5

The Rise of Positional Astronomy

D ESPITE his emphasis on planetary observations as a means of longitude determination, Halley made no observations on the eclipses of Jupiter's satellites; the tables of their motions which appear among his *Tabulae Astronomicae* being based upon observations by his young contemporary James Bradley (1692–1752), and eclipse tables of the first of the four known satellites on others by the latter's uncle James Pound (1669–1724). Bradley's determination of the mean motions of these bodies was the first to take cognisance of Roemer's discovery of the finite speed of light, the equation for which was fixed at the then accepted value of 14^m plus another of $3\frac{1}{2}^m$ for the effect of the eccentricity of Jupiter's orbit around the Sun. Giovanni Domenico Cassini had at first been reluctant to accept this brilliant discovery by Jean Picard's young Danish protégé since it negated a principal tenet of Descartes's philosophy. Even though it is evident from the official registers of the Paris Academy that most of the members were sympathetic to this view after Roemer's able defence of his hypothesis against Cassini's criticisms, the slowness with which it came to be accepted and applied by French astronomers was partially due to their high respect for Cassini's opinions. Another reason may have been the fear of offending him by pressing Roemer's view after the two had fallen out and the latter had gone back to Denmark. Nevertheless, Christian Huygens built his *Théorie de la Lumière* (Paris, 1690) upon the principle of the finite speed of light.

The discovery which finally put an end to this speculation by providing independent evidence for the finite speed of light, was made unintentionally by Bradley himself and immediately won him international fame because of its other implications, which we shall

91

be describing presently. On 17 December 1725, more from curiosity than any other motive, he was induced to observe the zenith distance of the bright star γ Draconis through the $3\frac{3}{4}$-inch aperture telescope of a 24-foot radius zenith sector at the private residence of Samuel Molyneux at Kew, near London. This remarkable instrument had just been constructed for that gentlemen by George Graham, the well-known instrument-maker who (only a few months earlier) had provided Halley with his 8-foot radius iron quadrant; and it was specially designed for detecting the presence of minute changes in the zenith distance of that particular star which could be ascribed to the long-sought influence of annual stellar parallax. γ Draconis, being a star which culminates very close to the zenith in the latitude of London and visible even by day, had been selected by Hooke for his enquiry into this phenomenon at Gresham College in 1669, using a 36-foot radius zenith sector, from which he had concluded that it exhibited an annual parallax of between 27" and 30". However, contemporary astronomers, including Bradley, were well aware of the poor construction and mounting of that instrument, and it was primarily on this account that his result was generally rejected. Flamsteed's own claim to have detected this phenomenon was criticized by Jacques Cassini, as has been mentioned earlier (cf. Chapter 2), and his observations ascribed to seasonal variations in atmospheric refraction; but this explanation was scarcely sufficient to account for the whole of the systematic discrepancy of 40" which was found. An annual change of over 1^m (or 15') in the difference of right ascension between Sirius and α Lyrae, which could not be accounted for either by parallax or refraction, had also been detected by Roemer in 1692–3.

It was generally appreciated that these anomalies could be explained only by a systematic series of observations extending over a period of at least one year with an instrument capable of measuring differences in zenith distance with as high an accuracy as possible. The possibility that such observations might provide the long-sought verification of Copernicus's postulate of the Earth's annual motion was an added stimulus to undertake such an investigation. Thus Samuel Molyneux, being an enthusiastic and wealthy amateur astronomer, had obtained the services of George Graham to construct and instal his sector—an operation requiring the insertion of a pivot in a chimney-stack and the cutting of holes in the floorboards of a ground-floor room of his house. The objective was in an upstairs room, and the viewing telescope was pointed through a hole made in the roof. Having initially adjusted a micrometer screw so that the star's apparent diurnal motion caused it to run along a cross-wire, subsequent changes in zenith distance could be measured by the amount by which that screw had to be turned to re-set the wire on the star. Molyneux made the final adjustments to his new instrument on 3 December 1725, and observed γ Draconis with it then and on three other occasions during the following fortnight; that is, before Bradley's visit. What surprised Bradley, when he put his eye to the instrument, was that the star appeared to cross the meridian a little farther to the south than where his friend had previously recorded it. There ought to have been no visible effect due to parallax at this time of year. A subsequent observation on 21 December confirmed the reality of this unexpected effect—the difference between his own and Molyneux's initial observation eighteen days previously, being 3·5″.

Their first doubts concerned the instrument itself, but their subsequent experiences with it convinced them of the stability of its suspension throughout very unfavourable spells of weather, and its consequent reliability for yielding the required relative differences in zenith distance to within ±0·5″. A second possibility was that the effect was produced by a 'nodding', or nutation, of the Earth's polar axis which Newton had predicted as one consequence of the Moon's variable perturbation on the Earth's annual orbit, but observations of a star in the constellation of Auriga (nicknamed 'anti-Draco') which crossed the meridian about twelve hours after γ Draconis revealed that it moved only 5″ northwards in the same time taken by the latter star to move 9·1″ southwards, which implied that this might be a contributory cause but not the entire cause of this behaviour. The fact that both stars passed within 0·1° of the zenith was in itself sufficient to rule out the possibility of refraction being responsible for the observed changes, and the mystery deepened when it appeared that the latter were dependent upon the stars' angular distances from the plane of the ecliptic.

In order to discover the law governing this effect, it therefore became desirable to extend these sensitive measurements to a wider range of celestial latitudes than the 8′ attainable with Molyneux's instrument. Thus Bradley got Graham to make an even more precise zenith sector of just over half the radius of Molyneux's and with a scale extending to $6\frac{1}{4}°$ on either side of the zenith. With this instrument, which he installed at his aunt's house in Wanstead (London) and which is today preserved in the old Royal Observatory at Greenwich, Bradley could observe 200 stars, twelve of these being bright enough to be seen at all seasons of the year. The results of a further year's observations at Wanstead between August 1727 and August 1728 established that the range in the variation of meridian zenith distance (or declination) of a star was proportional to the sine of its celestial latitude. Having considered and rejected the phenomena of nutation and refraction, and the possibility of changes in the direction of the plumb-line with respect to which the instrument was constantly rectified, Bradley finally conjectured "that all the Phenomena hitherto mentioned, proceeded from the progressive Motion of Light and the Earth's annual Motion in its Orbit".

This insight apparently occurred to Bradley during a pleasure-cruise on the River Thames, possibly in September 1728, when he perceived that every time the boat was turned to sail in the opposite direction, the weather-vane on the top of its mast shifted a little, as if there had been a slight change in the direction of the wind. However, after the sailors had assured him that this apparent change was actually due to the change in the motion of the boat, he immediately saw the relationship of this phenomenon to the problem that he had almost despaired of solving. The wind velocity he regarded as being analogous to that of light ($=c$, say) and the speed of the boat through the water to that of the Earth's orbital motion around the Sun ($=v$, say).

In his letter to Halley "giving an Account of a new discovered Motion of the Fix'd Stars", which was read on 9 and 16 January 1729 at successive weekly meetings of the Royal Society, Bradley summarizes his previous work and gives a geometrical demonstration of the effect which was soon to become known as stellar aberration (although the word itself was not introduced in that paper). His observations implied that $v : c = 1 : 10210$, hence that light is propagated from the Sun to the Earth in $8^m 12^s$, which lay comfortably between Roemer's own estimate of 11^m and Giovanni Domenico Cassini's of just over 7^m for the same distance based upon their independent observations of the eclipses of Jupiter's satellites. In view of Bradley's awareness of the superior accuracy of his own estimate, one might have expected him to have changed his equation for the finite speed of light which was ultimately published in Halley's posthumous tables. However, these had already been printed at this time and the expense involved in revising all the figures for the duration of the satellites' eclipses might have been the prohibitive factor.

Since the zenith sector observations had yielded variations only in the stars' declinations, Bradley, in his letter to Halley, makes no analysis of the apparent motions in right ascension. He refers, however, to the annual elliptical motion about a mean position resulting from combining the variations in a star's two equatorial coordinates, which reveals that he was fully aware of the existence of this related phenomenon. Eustachius Manfredi would appear to have been the first person to verify that the anomalies observed in right ascension were actually conformable to Bradley's hypothesis, but such comparisons were seriously affected by clock-errors. Indeed, such was the soundness of Bradley's interpretation, that clock-errors could be more reliably determined from his observations of changes in meridian zenith distances (or declinations) than vice versa. The main significance of his discovery, as far as his contemporaries were concerned, was that it provided the first experimental proof of the Earth's orbital motion and spelt a death-blow to current anti-Copernican world views. Ironically, this was just what Molyneux himself, who unfortunately died before Bradley had succeeded in interpreting the phenomenon which they had both been observing, had hoped to do (albeit by discovering a different phenomenon!). From the practical point of view, this important research established a higher standard of precision in astronomical observation.

Yet Bradley was still not satisfied. He continued to suspect a retrogression of the equinoctial points with an associated periodic inclination of the Earth's polar axis towards the ecliptic and back again, as Newton (in Book 3, Prop. 21, Theorem 17 of his *Principia*) had predicted would arise from the variable gravitational attraction of the Moon upon a spheroidal Earth. He well knew that by establishing the existence of such a phenomenon he could in effect indirectly be verifying the other Copernican postulate of the Earth's axial rotation since, according to Newton's physics, this was fundamentally responsible for the Earth's spheroidicity. Its precise determination would therefore provide an indirect means of calculating that quantity, and settle the dispute between Newtonians and the followers of Jacques Cassini whose trigonometrical measurements of the length of the degree of longitude and latitude in the neighbourhood of Paris had led him to believe that the Earth was elongated rather than flattened at the poles. The problem this time required even more perseverance than before, since this nutation was linked to that of the Moon's nodes and had a period of about 18⅔ years.

Nothing daunted by the prospect of extending his zenith sector observations by so many more years and perhaps inspired to emulate Halley's example, Bradley undertook this task and finally disclosed his long-standing conviction to the world in the form of "A Letter to the Right Honourable George Earl of Macclesfield concerning an apparent Motion observed in some of the fixed Stars", read at a meeting of the Royal Society on 14 February 1747. He stressed here that his success in quantifying the effects of aberration and nutation, and separating the latter from the accepted value of 50″ per annum for the precessional constant, was due primarily to the advice and assistance which he had received from George Graham. The Paris Academy expedition to Tornea in Lapland, led by Pierre de Maupertius with whom Bradley had been in correspondence concerning nutation shortly before his departure, had also employed a zenith sector by that same instrument-maker similar to the one at Wanstead. The results of this and Pierre Bouguer's independent expedition to Peru, favoured Colin Maclaurin's elegant geometrical verification of Newton's opinion that the Earth must be flattened at the poles if in equilibrium under gravity and the centrifugal force of the diurnal rotation. The trigonometrically deduced value for this oblateness was about twice that which the application of the principles of Newton's mechanics led one to anticipate, but Alexis Claude Clairaut, who had accompanied Maupertius's geodetic expedition, had been able in 1743 to account for this apparent discrepancy by dispensing with the artificial assumption that the Earth is homogeneous throughout. The substitution of values for the gravitational acceleration at different terrestrial latitudes into a mathematical formulation known as Clairaut's Theorem, now enabled the precise degree of ellipticity—hence the oblateness—to be found. However, it was Clairaut's fellow-countryman and scientific rival Jean D'Alembert who was the first to provide in 1749 a rigorous analytical demonstration of Newton's theory of the precession of the equinoxes and Bradley's hypothesis of the nutation of the poles.

Thus Bradley's twenty years of devoted labours at Wanstead, and his clear insights into the causes of the phenomena which he so

Fig. 7. SIR JONAS MOORE (1610–1679)
Surveyor General of HM Ordnance

Fig. 8. JOHN FLAMSTEED (1646–1719)
First Astronomer Royal from 1675 to 1719

Fig. 9. SIR ISAAC NEWTON (1642–1727)
President of the Royal Society

Fig. 10. EDMOND HALLEY (1656–1742)
Second Astronomer Royal from 1720 to 1742

Fig. 11. JAMES BRADLEY (1693–1762)
Third Astronomer Royal from 1742 to 1762

Fig. 12. NATHANIEL BLISS (1700–1764)
Fourth Astronomer Royal from 1762 to 1764

Fig. 13. NEVIL MASKELYNE (1732–1811)
Fifth Astronomer Royal from 1765 to 1811

Fig. 14. JOHN POND (1767–1836)
Sixth Astronomer Royal from 1811 to 1835

patiently and systematically investigated, were largely instrumental in providing the much desired experimental proof of the fundamental axioms of Copernican astronomy (viz. the Earth's annual and diurnal motions) and the principle of universal gravitation on which Newton's System of the World was based. But, as the twentieth-century historian of science George Sarton has already emphasized, even if he had made neither discovery,

> "he would still have been remembered as one of the very best experimentalists of his time, and one of the founders of modern astronomical observation. He accumulated scientific data whose quality is as remarkable as their quantity. Their publication and reduction occupied many mathematicians for a long period after his death."[1]

The allusion here is doubtlessly meant to include August Busch's reduction of Bradley's Kew and Wanstead observations, published in Oxford in 1838. In particular, however, it refers to the painstaking analyses by Friedrich Wilhelm Bessel and Arthur Auwers on approximately 60,000 observations of over 3200 stars made by Bradley during his period of office as Third Astronomer Royal from the year of Halley's death (1742) until his own death twenty years later; and very probably also to Francis Baily's thorough review of a whole host of existing star catalogues (including Bradley's) in *The Catalogue of Stars of the British Association for the Advancement of Science* (London, 1845). Baily's well-qualified assessment of Bradley's role at the Greenwich Observatory on p. 74 of the preface to that very important work is worth quoting, since it relates the latter's endeavours to the pioneering efforts of Flamsteed.

> "Bradley's labors at The Royal Observatory, in this department of the science, consist almost wholly of a re-observation of the stars in Flamsteed's catalogue. He caused these stars to be reduced to the year 1744, and arranged in the order of right ascension, as a sort of working catalogue for his own use; which book still exists in the library of the Royal Observatory. Very few other stars have been observed by Bradley, except such as occasionally entered the field of his telescope whilst he was watching for those of Flamsteed. We are thus indebted to Flamsteed for the subsequent labours of Bradley: for had not Flamsteed led the way, there is much doubt whether Bradley ... would have pursued a similar independent course."[2]

E

Certainly, this is one way of looking at Bradley's contributions to the development of positional astronomy; but the picture presented is rather unfair to him. To begin with, one should ask oneself whether it was better for Bradley to seek to introduce a greater degree of precision in the coordinates of the brighter stars already approximately determined by his illustrious predecessor, or to extend the observations to fainter stars? From the standpoint of improving navigational practice—the aim for which the Royal Observatory had been erected—surely the former course, which Bradley followed, was the correct one! But there is a deeper motivation in his policy, which Baily seems not to have appreciated. Bradley was aware, as a consequence of his own earlier discoveries described above, that every star position had to be carefully corrected for the effects of precession, aberration, and nutation before any significance could be attached to the differences which he knew to exist in the *relative* angular separations between pairs of stars. If the reality of this phenomenon could be first established, the question of whether these changes were reflections of the motion of our solar system through space, or arose from the stars' own individual motions, could be considered. Intercomparisons between pairs of bright stars, which could generally be regarded as being nearer to us than the fainter ones, would be most likely to exhibit such motions (whether apparent or real). Bradley reveals his conscious awareness of these matters by explicitly mentioning them near the end of his paper on nutation. Thus there can be little doubt that he had a strong theoretical motive for concentrating his attention on improving the degree of accuracy with which the principal stars were determined. Moreover, he knew that in order to do so, the variable effects of refraction would have to be estimated with a great deal of care, with the aid of only the most reliable instruments; or even rendered unnecessary by selecting the stars culminating very close to the zenith with which he was already thoroughly familiar by the time he took up his appointment at Greenwich in June 1742.

The state of the instruments at the Royal Observatory was, however, a cause for concern. The 8-foot radius iron quadrant which had been made by Jonathan Sisson under George Graham's supervision and which Bradley himself had helped the latter to adjust and mount in September 1726, was jammed against the roof of Halley's Quadrant House. Flamsteed's Sextant House, to the south of that building, had become a pigeon loft; and a well telescope, which had never been much used, was blocked up. Several very necessary, and very material, alterations had to be made to Halley's mural quadrant and to his 5½-foot focus transit instrument. Thus before commencing a regular series of meridian observations at the beginning of the following year, Bradley balanced the telescope of the latter instrument and adjusted its horizontal axis, invented an apparatus for illuminating the wires of the micrometer eyepiece in all positions in the focal plane, and fixed a horary wire on each side of the middle wire. He determined the position and direction of both instruments with very great accuracy. With these improvements he observed the positions of the Sun, planets and fixed stars with a degree of accuracy hitherto unknown. The major source of uncertainty lay in the necessity of applying a mean correction for the effect of refraction, owing to the lack of precise knowledge of the state of the Earth's atmosphere. Since, however, a considerable amount of force had sometimes to be applied to the viewing telescope, due to the warping of the timber in the roof of the room containing the quadrant which caused the lead counter-weight to rub against it, not all observations were equally accurate. Instrumental errors were also introduced on account of the difficulty in adjusting the line of collimation of the transit telescope.

Accordingly, Bradley (supported by the members of the Royal Society) presented a memorial to George III asking for better instruments and a better building to house them in. As a result, John Bird was commissioned to construct a new transit instrument, another 8-foot radius quadrant—this time made of brass—and such other instruments as might contribute to the advancement of astronomical knowledge. Both the transit instrument and quadrant, together with Bradley's own 12½-foot radius zenith sector from his Wanstead observatory, were suitably mounted as soon as the new building at the Royal Observatory was completed in 1750. A 40-inch radius movable quadrant also made by John Bird, and an excellent clock made by John Shelton under George Graham's direction, were provided at the same time and placed in the transit room. These were the instruments with which Bradley observed assiduously until his death in 1762.

Dr Bradley's observations, made in the course of the twenty years between 1742 and 1762, consisted all together of thirteen folio volumes (containing 931 pages), and two in quarto. After his death they were removed from the Royal Observatory by the executors of his estate, who refused to hand them over to the Royal Society, since they "could not conceive to be consonant with reason and equity"[3] that such an extensive and accurate collection of data should be given up in consideration of the small and inadequate salary that he received. A pension granted to Bradley in 1752 by King George II had not been in lieu of an increase in salary, but had been a gratuitous acknowledgment of his personal merit, and of certain services performed by him in an unofficial capacity. There was no precedent to the Royal Society's request for these papers, since Flamsteed had printed a considerable part of his own observations during his lifetime for his personal emolument, and Halley's daughter (Mrs Catherine Price) had received compensation for her father's observations after his death.

For these reasons, when, in 1765 the Board of Longitude asked the Principal Secretary of State to make a similar request on their behalf, he was doubtful whether he had the authority to do so; whereupon it was decided to put the case before H.M. Attorney and Solicitor-General (Messrs Yorke and De Grey). But the legal complications proved to be greater than anticipated, and it was therefore decided to suspend proceedings until Bradley's daughter (his only child) should come of age in January 1767, when she should be asked to hand over the desired papers.

Unfortunately, it appeared that Miss Bradley had not been aware of the fact that the latter were her sole property, for on coming of age she handed them over to her uncle, the Rev. Samuel Peach, who flatly refused to give them up "without a very valuable consideration".[4] The matter was eventually referred to the King's Advocate, Attorney, and Solicitor-General, who recommended that the observations be recovered by his issuing "an information in the name of His Majesty's Attorney-General in the Court of Exchequer",[5] to which the King gave his assent.

Thomas Hornsby, D.D., who, as Savilian Professor of Astronomy at the University of Oxford, was one of the commissioners of longitude, communicated this decision to the late Samuel Peach's elder son, John, who had now inherited his uncle's observations. But John Peach was just as stubborn as his father had been, and refused to part with the papers, so the secretary of the Board of Longitude was instructed to apply to the Solicitor of the Treasury for a copy of an information issued against John Peach, which was to be communicated to the late Dr Bradley's representatives and the matter then brought into court. While this was going on, Mrs Elizabeth Tew of Islington entrusted, under seal, to the secretary of the Royal Society, Abram Robertson, Flamsteed's original observations which had been made between 1670 and 1719 "in many Volumes in Folio and Quarto bound and unbound together with a great number of original Letters from the most Learned men of his Age",[6] and surrendered them to the Board of Longitude for the sum of £100.

Towards the end of 1772, John Peach sent a memorial to the Board asking whether the law-suit would be called off, but the commissioners resolved that it be continued. They were already committed to adopting this attitude, in view of the encouragement given in the interim to Mr Francis, who was employed under the direction of Mr Nuthall, one of the solicitors of the Treasury, to carry on the prosecution against the representatives of the late Dr Bradley without delay. On being informed of this resolution, Peach is reported as having retorted that as he thought he had some property in the papers he could not agree to give them up without the certainty of a proper gratuity. Nevertheless, just over three years later, when the matter was about to be taken to court, his younger brother the Rev. Samuel Peach (Junior) who had become the legal heir to the observations on his marriage to Miss Bradley, presented them in 1776 to Lord North (later, Earl of Guildford), Chancellor of the University of Oxford, who, being unaware of the foregoing events, unwittingly presented them to that university on condition that they would be printed and published by the Clarendon Press. It was then generally believed that the Crown could no longer claim the right to these papers; consequently, the law-suit was terminated later that same year, since both the Board of Longitude and the Royal Society were satisfied with the assurance that the manuscripts were at least in the hands of printers. However, when over fifteen more years had passed by and Bradley's observations were still in the press, everyone concerned with the promotion of navigation and astronomy had become impatient at the undue delay. Sir Joseph Banks, President of the Royal Society and an eminent member of the Board of Longitude, wrote to the Vice-Chancellor of the University of Oxford regarding this matter, which elicited the following reply through the Rev. Samuel Forster, Register to the Delegates of the Clarendon Press:

"Sir   Oxford June 12, 1791

I received your Letter; and duly communicated it to the Delegates of the Oxford Press at their usual meeting immediately before the expiration of Term. From whom I am instructed to say that they are not conscious of any unnecessary delay respecting the publication of Dr Bradley's Papers: But that their Astronomical Professor [Hornsby] into whose hands they were originally put, has given every requisite attention to them, and made as much progress as his health and the time would admit of—That a considerable part of them is already printed, and in the most handsome and useful manner—That the only impediment they have labouren under, has been the unforeseen Illness of Dr Hornsby, which (tho' not so severe as wholly to interrupt his Academical Lectures) has for some time past retarded him in the prosecution of the Work ... The Board of Delegates conclude that the Council of the Royal Society will agree with them, that under such circumstances the utmost respect and even delicacy is due to the Professor, and that the Papers should not be hastily taken out of his hands nor himself too much pressed upon the subject."[7]

The Council of the Royal Society, however, did not agree that Professor Hornsby's ill-health was a sufficient justification for such a long delay, for they considered that a capable astronomer could have been found with the ability to execute the whole of the work in less than six years. A formal communication was then sent, only a few months later, to the Rev. Dr Cook, Vice-Chancellor of the University of Oxford, requesting that a fair copy of these observations, which had already been printed, and new pages as soon as they would be ready, should be sent to the Astronomer Royal in order to bridge the "chasm of twenty years" (from 1742 to 1762) in the series that had been made from the time of the foundation of the Royal Observatory in 1676 until then (1791).[8] The delegates of the Clarendon Press took this letter into consideration, yet did not comply with that request, merely assuring the Board that despite unavoidable delays the work of printing and publishing Dr Bradley's observations was finally well under way. A year later, Professor Hornsby also added his assurance (as editor) that the first volume to these papers would be published early in 1794; but this estimate failed to take into account the most significant factor, his own continued ill-health, which contributed appreciably to a further delay.

The Board of Longitude then appointed a committee to hasten the publication of Bradley's observations; but their attempt to arrange a meeting with the Duke of Portland, Chancellor of the University of Oxford, for the purpose of discussing this matter, was unsuccessful. The Board's letter to the Duke did, however, have the effect of stimulating a further enquiry into the progress and state of the work, as a result of which the Board of Delegates again declared themselves to be satisfied that Dr Hornsby himself was not to blame for the delay and that there was

> "every reason to hope that the first volume will very soon be published, without haveing recourse to the disagreeable expedient of taking the Papers out of Dr Hornsby's hands; which, from a variety of considerations, they [the Delegates] are unwilling to adopt, and conceive it might eventually tend to the hindrance, rather than the furtherance of the publication".[9]

At Professor Hornsby's own suggestion, each commissioner of longitude was issued, towards the end of that year, with two printed copies of the Board's minutes and other relevant papers in order that the more recently appointed members should be fully acquainted with the entire proceedings over the whole thirty-year period while the problem of the recovery of Dr Bradley's observations had occupied the Board's attention. Copies were also sent to Oxford common rooms, to the delegates of the Clarendon Press, and to a number of other interested parties.

The first volume of Bradley's "Astronomical Observations . . ." was finally published in 1798. It includes preliminary tables of aberration in the equatorial coordinates of sixteen stars (at ten-day intervals for one year), a table of astronomical refractions, a catalogue of 389 fixed stars from the 1st to 7th magnitudes, and longitudes and latitudes of the Moon deduced from observations made between 13 September 1750 and 2 November 1760 (inclusive). Then follow tables giving the times of "Observed Transits of the Sun, Planets, and Fixed Stars over the Meridian" from September 1750 to the end of December 1755, these being estimated to the nearest one-third of a second; "Meridional Distance of the Sun, Planets, and Fixed Stars, from the Zenith, Southward" from September 1750 to December 1758, giving the barometric and thermometric readings at the time of observation; "Observations made with the Zenith Sector" from December 1749 to September 1760; "Daily Rates of the Clock, used in Calculation" from 2 September 1750 to 31 December 1755; and "Apparent Right Ascensions" also between these latter dates.

The second volume, published seven years later under the editorship of Abram Robertson, who had been appointed Hornsby's successor as Savilian Professor of Astronomy at Oxford, contains similar observations made by Dr Bradley between 1756 and 1762. Robertson remarks in the preface that he had found it necessary to reduce to degrees, minutes, and seconds about 3000 observations made with the exterior divisions of Bird's quadrant, which had not been done by Bradley in his later years. In order to provide a connecting link between Bradley's observations and those of his successors, the Board of Longitude presented to the delegates of the Clarendon Press, the interim manuscript observations made by the fourth Astronomer Royal, Nathaniel Bliss, and his assistant Charles Green who had continued the observations at the Royal Observatory from the day of Bliss's death until 15 March 1765. These had been purchased by the Board almost thirty-six years previously. The offer was accompanied by another from Bliss's replacement, the Rev. Nevil Maskelyne, of certain calculations made from these manuscripts while they had been in his possession, including daily rates of the transit clock, mean times of the Moon's meridian passage (with longitudes and latitude, etc.), and the apparent right ascensions of the planets. All these data were printed in his second volume as an addendum to Bradley's own work.

Some years later, Professor Friedrich Wilhelm Bessel of Königsberg set to work on the complicated task of reducing these observed star positions from their apparent to their true values, and in 1817 received financial encouragement from the Board of Longitude to assist in the publication of his results. Born in Minden in 1784, Bessel had long been interested in practical navigation; having begun his working career as an apprentice to the famous mercantile firm of Kulenkamp of Bremen, where he gained proficiency in calculation and commercial accounting. While training to be a cargo officer on a merchant ship, he learned the technique of determining longitude at sea from measurements of lunar distances, and studied the underlying mathematical theory until he was fully conversant with that method. He subsequently extended his astronomical interests to the calculation of the orbit of Halley's comet using the observation made by Thomas Harriot in 1607 which he found in a supplementary volume of the *Berliner astronomisches Jahrbuch*, applying the computational principles of Heinrich Wilhelm Olbers. This induced him to present the results of his calculations to Olbers, who was greatly impressed with their close agreement to the elliptical elements calculated by Halley and encouraged him to publish an impressive article on the subject. This marked the turning point of Bessel's career.

On Olbers' recommendation, Bessel left Kulenkamp's before completing his apprenticeship to begin work as an assistant to Johann Hieronymus Schröter at the latter's private observatory at Lilienthal near Bremen; and it was here that he acquired practical experience observing comets and planets. Observationally, he paid special attention to the system of Saturn, while building up a thorough knowledge of celestial mechanics and making further calculations for the determination of cometary orbits.

Two years after the publication of the second volume of Bradley's observations (1805), Olbers encouraged him to apply his expertise to the reduction of the positions of the 3222 stars which had been observed at Greenwich between 1750 and 1762. At the outset of this investigation Bessel recognized the need for a reliable reference system for the measurement of star places and the determination of accurate coordinates for the earliest possible epoch, in order to free these from the effects of the Earth's motion and site of the observation. For this purpose, precise corrections for precession, aberration, and nutation, and the latitude of the Greenwich Observatory for a mean epoch (1755.00), had all to be obtained. He was able to make a comparison of Bradley's star positions with those in Giuseppe Piazzi's *Praecipuarum stellarum inerrantium positiones mediae*, etc., based on observations made in the Palermo Observatory between 1792 and 1802, and subsequently of the continuation of that catalogue for observations up to 1813, for determining precession from proper motions. The outcome of these researches is contained in Bessel's *Fundamenta astronomiae* (1818), a work which is generally considered to have been one of his greatest contributions to this science. When referring to Francis Baily's *New Tables for facilitating the computation of Precession, Aberration and Nutation of 2881 principal fixed Stars, together with a Catalogue of the Same reduced to January 1, 1830* (London, 1827), Sir John Herschel praises Bessel's achievement in the following terms:

> "—the distinguishing characteristic of this work is the adoption throughout of Professor Bessel's capital improvement in the system of applying the corrections, by arranging the formulae in such a manner that all that is peculiar to each star, and permanent in

magnitude, shall stand distinctly separated from all that is ephemeral, or varying from day to day; and that, in such a manner that a short ephemeral table, capable of being compressed into a single page, shall serve not only for these stars, but for every star in the heavens. The convenience of this method, the brevity it introduces into the computations, the distinctness it gives to all the process of reduction, requiring neither thought nor memory on the computer's part, give it an incalculable advantage over every other. To reduce any observation, no other book need be opened. The work occupies four lines, and is done in half that number of minutes."[10]

Bessel's later *Tabulae Regiomontanae* (1830) contains the mean and apparent positions of thirty-eight stars for interim observations made at Greenwich after Bradley's death on all but two of those stars, and his own observations from 1813 onwards at the Königsberg Observatory of which he had been appointed director by Friedrich Wilhelm III of Prussia before its construction was begun. The star positions given in these tables constitute the first modern reference system for the measurement of the positions of the Sun, Moon planets, and stars. With their aid, all such observations made since 1750 at the Greenwich Observatory could be reduced, and these data could be used for developing the theories of planetary orbits. An important outcome of Bessel's study of proper motions was his recognition of the fact that the fastest-moving stars, not the brightest, are nearest to us, followed by his consequent discovery of the parallax of the star which Flamsteed had designated as 61 Cygni. The value assigned by him to this quantity in the *Astronomische Nachrichten* of 1838, was $0.314'' \pm 0.020''$ (as compared with the modern accepted value of $0.292'' \pm 0.0045''$). The minuteness of this parallax explains why earlier astronomers such as Flamsteed himself had sought for the phenomenon in vain. Now, however, it was evident that the distance to the innermost regions of the stellar universe could be found by direct trigonometrical measurements. It was appropriate that Bessel's work should thereby have provided independent confirmation of Bradley's discovery over a century earlier that the Earth is indeed revolving annually around the Sun as Copernicus and a host of other astronomers after him had surmised but had been unable to establish by experimental means.

The tradition for accurate postitional astronomy established by Bessel at Königsberg was maintained by Arthur Auwers who, from 1865 to 1883, undertook a completely new reduction of Bradley's observations, including many hitherto unpublished earlier ones between 1743 and 1750 which had been found after Bessel's work had been completed. The final result was the publication of three volumes of Bradley's observations (Leipzig, 1912–14), which can justly be regarded as the foundation of all modern star positions and proper motions and constitute proof that his influence was still being felt well into the present century.

NOTES

1. Sarton, 1931, p. 236.
2. Baily, 1845, p. 74.
3. Hornsby (Ed.), preface.
4. R.G.O. MSS. 533, Board of Longitude Confirmed Minutes, 5 (4 March 1769), p. 89.
5. *Ibid.* (13 January 1770), pp. 93–94.
6. *Ibid.* (28 November 1771), p. 105.
7. R.G.O. MSS. 532, Board of Longitude, 4, pp. 247–248.
8. *Ibid.*, p. 268. This letter was sent in Maskelyne's name, and dated 22 December, 1791.
9. *Ibid.*, p. 271. This letter was addressed to the Duke of Portland, and dated 27 February 1795.
10. Quoted in Baily, 1845, p. 8.

6

Outside Influences

BRADLEY's labours at Greenwich, important though they undoubtedly were, ought not to be regarded as unique. The
nineteenth-century British historians of astronomy, Agnes Clarke
and Stephen Rigaud, who eulogize his achievements and rank him
as the founder of modern observational astronomy, are inclined to
underestimate the fundamental improvements of the observing
techniques of his eminent continental contemporaries Nicolas Louis
de Lacaille and Tobias Mayer. The respective contributions of these
men to the lunar distances method of longitude determination at
sea were also destined to have an enormous influence upon the
subsequent history of the Greenwich Observatory, for reasons
which will now be expounded. Joseph Delambre's study of the
relative merits of the trio in *Histoire de l'astronomie au XVIIIe
siècle* (Paris, 1827), composed after he had completed a review of the
progress of astronomy in ancient and medieval times and written
a further volume on the advances made by seventeenth-century
astronomers, led him to swallow his national pride by placing Mayer
first, followed by Lacaille, and finally Bradley. It is not necessary to
repeat his rather technical assessment here. Suffice it to say that while
praising Bradley as "the author of the two most brilliant and useful
discoveries of the [eighteenth] century",[1] he criticizes him for failing
to make proper allowances for instrumental corrections when
seeking to obtain the *absolute* values of stellar coordinates; in particular, for his neglect of collimation errors in his mural quadrant,
which Bessel had shown to be significant. Aberration and nutation
had involved the measurement of *relative* differences only, where a
high degree of precision could more easily be achieved.

In a *Mémoire* written in 1749, Lacaille remarks that it had been
Bradley's establishment of the value of nutation which first
prompted him to embark upon the construction of a new set of

solar tables. In order to determine the elements of his solar theory, Lacaille compared his own observations at the Mazarine College in Paris with those of Regiomontanus's pupil Bernhard Walther at Nuremberg and found that

(1)   the obliquity of the ecliptic was diminishing at about 20″ per century (not constant, as Flamsteed had believed)

(2)   the longitude of the solar apogee was increasing by almost 14″ per year in excess of the effect due to precession which could be more accurately evaluated now that the nutational constant was known

(3)   the mean length of the tropical year, defined as the time taken by the Sun to return to the same equinox, was $365^d\,5^h\,48^m\,46^s$ over the 250-year interval between the two sets of data.

Lacaille also deduced that the eccentricity of the Earth's orbit had remained constant during that time, at a little under $\frac{1}{60}$. He made over fifty comparisons between the position of the Sun and that of Vega, each involving numerous observations extending over several days and the correction of that bright star's places for the effects of precession, nutation, and aberration. His claim to be able to estimate the differences in right ascension between the Sun and Vega to within ±5″ was slightly exaggerated, since the calculated longitudes were in error by over twice that amount.

Irregularities in the Sun's apparent motion made Lacaille suspect that the Moon's gravitational attraction was perturbing the Earth's elliptical orbit, the reality of which he was able to confirm by an analysis of fifteen carefully selected solar observations. His expedition to the Cape of Good Hope in 1750–52 enabled him to observe the Sun near the perihelion where its apparent motion is most rapid and where it is least accurately observed from the northern hemisphere. Thus he was able to refine the observational basis of his own solar theory at the same time as Alexis Claude Clairaut and Leonhard Euler were improving the lunar theory by developing it analytically and taking account of the Newtonian gravitational attractions upon the Earth by the Moon, Jupiter and Venus. This naturally introduced further complications into the apparent motions of the Sun. He also had access to late fourteenth-century measurements of the Sun's solstitial altitudes made with a 40-foot gnomon at Pekin, communicated to Joseph Nicolas Delisle by Père Antoine Gaubil, a Jesuit missionary in China. These indicated a diminution of 47″ per century in the obliquity of the ecliptic, which was consistent with new information supplied to him by Tobias Mayer regarding the site of Walther's observatory and necessitated a correction being made to his former estimate of 20″. He therefore re-determined the elements of the Sun's apparent orbit with the aid of all the above-mentioned improvements and found that his predictions were invariably correct to within ±15″ whereas those based on Halley's and Jacques Cassini's solar tables were uncertain to ±2′. Lacaille's *Tabulae Solares* of 1758 is of great historical importance in being the first to take account of Bradley's discoveries and the gravitational perturbations of the Moon and planets.

There was a close connection between Lacaille's reformation of the solar theory and his plans for constructing a more accurate star catalogue. The conventional point of reference for the measurement of the longitudes of all heavenly bodies is the Vernal Equinox or point of intersection of the ecliptic and celestial equator. However, as early as 1742, Lacaille had informed the members of the Paris Academy of his intention to introduce the new procedure of dividing the heavens into zones parallel to the celestial equator and of referring the stars in each zone to one or more principal stars whose coordinates he would determine fundamentally. The constant interruption of this observing programme by bad weather was partly responsible for his decision to go to the Cape, where he could be assured of clear skies and a temperate climate. The observations of the Southern skies by the youthful Edmond Halley in St Helena were now also in great need of improvement, and the establishment of a temporary observing station so far to the south on a meridian passing right through the heart of western Europe seemed a good way of finding the trigonometrical parallaxes of the Sun, Moon, and planets. The length of a degree of that meridian, and the length of the seconds pendulum at the Cape were both worth determining to see whether the distance measured and the value derived for the gravitational acceleration conformed to Clairaut's theoretical value for the Earth's oblateness; it was also desirable that the longitude of the Cape should be more precisely established.

Using a 26-inch focus telescope of ½-inch aperture attached to his 3-foot radius iron quadrant, Lacaille observed 10,000 stars—selecting about 1900 of them for inclusion in a catalogue—in the position of the sky between the south celestial pole and the Tropic of Capricorn, which he divided into 25 roughly equal zones parallel to the equator. In the focal plane of the eyepiece was a reticule consisting of four equal strips of metal connected to form a rhombus or lozenge-shaped outline in the observer's field of view. The longer diagonal of the rhombus was placed vertically and coincided with the meridian. The *mean* of the times at which the star entered and left the clear space enclosed by the strips gave the instant when the star crossed the meridian (i.e. its right ascension), whereas their *difference* gave the length of area traversed by the star within the rhombus and hence the difference in its declination from that defined by the horizontal diagonal. He judged the star positions found in this way to be reliable to within $\pm 30''$.

His Paris observations of the northern stars had been made using a different technique, which he also employed at the Cape for finding the coordinates of the reference stars for each zone: it involved timing the transit of a star by averaging the times at which it was at equal altitudes east and west of the meridian. All he required for this purpose was his 3-foot radius quadrant and a good clock (which he set to sidereal time); and, to attain greater precision, he customarily took about a dozen altitudes on each side of the meridian. An advantage of this method was that only the *relative* altitudes had to be known, but the sky had to be clear of clouds at two instants rather than one before the transit time could be found in this manner. The declinations were deduced from the meridian zenith distances of the stars measured with a 6-foot radius sector, knowing the latitude of his observatory, making due allowances for the effects of refraction, aberration, and nutation, and reducing his results from the apparent places to the mean places for the epoch 1750.0. He published these data for 307 stars in both hemispheres, in his *Éphémérides des mouvemens célestes 1755–1765* (Paris, 1755), and subsequently for 398 stars in the *Astronomiae Fundamenta* (Paris, 1757). The latter volume, which fully merits its title, includes 144 places of the Sun which served as the basis for his *Tabulae Solares* of 1758. Later editions of Lacaille's star catalogues were published by Francis Baily (1833) and C. R. Powalsky (1882).

The quantitative determination of the atmospheric refraction presented one of the most serious problems of Lacaille's reductions. In 1755 he published an important memoir on this subject based upon an original procedure involving the skilful combination of observations of the same stars at Paris and the Cape. He reduced the values which he obtained to their mean values under standard conditions of pressure and temperature, assuming the effect to be proportional to the density of the air and hence to the barometric pressure. He used an empirical rule communicated to him by Tobias Mayer for making the temperature corrections, but an error in the graduation of the sextant employed for this investigation caused him to over-estimate the effect of this phenomenon.

Notwithstanding the importance of the above-mentioned results of Lacaille's expedition to the Cape for the science of positional astronomy, the Paris Academy regarded the most useful part of it to have been his experiments on the determination of longitude at sea throughout the eight-month period of his double voyage. Lacaille's memoir of 1759 on the history and theory of this problem recommended the calculation at four-hourly intervals in advance of a voyage, of the Moon's distances from seven or eight bright stars distributed around the zodiac. These could be compared with observations during the voyage, using a Hadley quadrant, of the Moon's illuminated limb from the more conveniently situated of these stars. The Moon's horizontal parallax at noon of each day, and the transit times of these stars, were also to be tabulated; and the appropriate corrections applied for refraction, the lunar parallax, and altitude changes during the ten minutes or so of the observation itself. Lacaille devised a graphical method to facilitate these corrections, which was subsequently included among the directions to navigators published in the *Connaissance des Temps* for 1761 (Paris, 1759). Thanks to the high accuracy promised by the new lunar theories being developed by Clairaut, Euler, and Mayer, Lacaille estimated that within forty-five minutes of making these observations, the longitude at sea could be determined to 1° or less. In 1760 he published the *Nouveau Traité de Navigation* of his late friend Pierre Bouguer, this being but one of several editorial and computational projects undertaken by him to promote navigation and the related science of cartography.

The epoch-making contributions of Mayer to the same area of scientific endeavour are of even greater significance for the history of the Greenwich Observatory because they caused Bradley to become deeply involved in the problem of longitude determination at sea and were directly responsible for the foundation of the *Nautical Almanac*. The annual publication of this important aid to celestial navigation from the end of 1766 onwards has continued to be a major facet of the work carried out at the Royal Observatory ever since. The work which heralded Mayer's arrival on the international scientific scene was his "New Tables of the Motions of the Sun and Moon", published in Latin in the second volume of transactions (or *Commentarii*) of the Göttingen Scientific Society early in 1753. These tables were the outcome of researches on the lunar theory begun with the intention of improving a lunar map by investigating the quantitative allowances that had to be made for the phenomenon of the Moon's libration around its axis; the map itself being to assist the lunar eclipse method of finding longitude differences between pairs of land-based observing stations by removing existing ambiguities in the identification of the Moon's most prominent topographical markings. Unfortunately, Mayer gave no indication in this publication of how he had derived his formulae, yet claimed an accuracy of within $\pm 2'$ in prediction which was appreciably in excess of that found by both Clairaut and Euler whose respective lunar theories had just been published by the St Petersburg and Berlin Academies of Science. His conviction that the stage had now been reached where the accuracy of the lunar observations could not match that of the theory was borne out by subsequent comparisons which he made with the published observations of Flamsteed and an uninterrupted series of 139 lunar positions observed at Greenwich during the years 1743–45 by Bradley, communicated to him via Euler and Christopher Schumacher. Mayer published them as an appendix to a subsequent article in the third volume of the same transactions for 1754, where he stresses the usefulness of his tables as a basis for the discovery of longitude at sea.

Although his theory differed by as much as $\pm 4'$ from Flamsteed's observations it agreed to within $\pm 1'$ with those of Bradley. An independent verification of this exceptional degree of accuracy for the particular cases of the Moon's conjunctions and oppositions with the Sun was that its differences with respect to the Moon's celestial position derived from a systematic study of lunar and solar eclipses observed after the invention of the astronomical telescope and pendulum clock never exceeded $1' 10''$; and Mayer felt that even these comparatively small discrepancies could have been largely due to the variation in the eccentricity of the Earth's orbit. Since $\pm 1'$ corresponds approximately to an error of $\frac{1}{2}°$ or 30 nautical miles in terrestrial longitude at the Equator, Mayer appreciated that his tables promised to ensure the practicability of the method of lunar distances in oceanic navigation, an aim which Halley had previously tried in vain to achieve.

After considering how the phenomena of lunar occultations, and lunar and solar eclipses, could serve as two independent means of testing the accuracy of his lunar theory, Mayer recognized that the principal cause of the larger discrepancies found by Clairaut and Euler was the reflection, in the celestial coordinates of the Moon, of errors in observed star positions. The resulting *apparent* errors in the Moon's position, though not affecting the values of the coefficients in the lunar inequalities, *did* affect the value of its mean motion deduced from the measured times of lunar occultations. These errors could, however, be corrected with the aid of the eclipse observations. Mayer consequently made use of his own observations of the occultations of bright stars to obtain over 100 accurate mean values of the Moon's longitude to within $5''$ to $10''$, and simultaneously improved the internal consistency of his tables by taking account of a larger number of terms in the infinite series. He then used these results to calculate the positions of other zodiacal stars and thereby discovered errors in contemporary star catalogues which, if ignored, would have led one to conclude that *apparent* errors of several minutes of arc were inherent in the Moon's places derived from the observations of occultations. Thus Mayer was induced to devote the remaining few years of his life in Göttingen to the elimination of instrumental errors in a 6-foot radius brass mural quadrant by John Bird modelled on the 8-foot prototype being used by Bradley, which was employed for the measurement of star positions; a simple and accurate method of computing solar eclipses; the construction of his own new catalogue of zodiacal stars; and an investigation of the proper motion of stars. His lectures on these four astronomical topics to his local scientific society between 1756 and 1760 were published posthumously in Georg Christoph Lichtenberg's *Opera inedita Tobiae Mayeri*, I (Göttingen, 1775) and dedicated to the Hanoverian monarch George III whom the editor wished to impress.

The link between Mayer and Bradley was first forged towards the end of 1754 at the instigation of Johann David Michaelis, a noted oriental scholar and Secretary for Hanoverian affairs in Göttingen, after he had sounded out the reaction in England to Mayer's lunar tables, which Euler had been praising to his colleagues at the Berlin Academy of Science as "the most admirable masterpiece in theoretical astronomy".[2] Michaelis was conscious of the honour which a successful claim by Mayer for one of the parliamentary bounties offered under the terms of the 1714 Act would bring to Göttingen University, and solicited the cooperation of his diplomatic counterpart in London, William Philip Best, in obtaining an informed judgment on this matter. Michaelis asked Best to forward a copy of the third volume of the Göttingen transactions (containing Mayer's later article with the above-mentioned lunar observations by Bradley appended to it) to George, second Earl of Macclesfield and President of the Royal Society, in the hope that he would bring it to people's attention. Best assured Michaelis that the English were already well aware of Mayer's talents, and quoted the Earl as having said that all the local astronomers were praising the tables published in the second volume of the Göttingen transactions as being the most accurate in existence. On the other hand, all were still very doubtful as to whether their application to the problem of longitude determination at sea would be practicable, mainly on account of the observational difficulties associated with the method.

Thus Michaelis encouraged Mayer to present the part of a lecture outlining the precepts for finding longitude *at sea* with the aid of his tables, read at a meeting of the Göttingen Scientific Society on 12 October 1754, in the form of a memorial to the British Admiralty together with a copy of the tables themselves and a covering letter requesting that these be considered for one of the longitude prizes. These papers were sent via the usual diplomatic channels of Michaelis, Gerlach Adolph von Münchhausen in Hanover, and Best to the First Lord of the Admiralty (Lord Anson); but before being dispatched on the final phase of its journey, Mayer's "Methodus Longitudinum Promota" was taken by Best at the Earl of Macclesfield's suggestion to Bradley in order to get his qualified opinion of its merits. In November 1754, Best and Bradley met at the Earl's home at Shirburn Castle to discuss this matter. Apparently Bradley spoke highly of Mayer's tables, which he agreed ought to be presented to the Admiralty in order that they might be tested. He thought that the adequacy of Mayer's Repeating Circle, a reflecting instrument proposed by its inventor as an alternative to the Hadley quadrant for observations at sea, need not be considered at that stage. Since Mayer had claimed in his 1754 article that he had improved upon the tables published one year previously, Bradley thought that he ought to produce his further calculations and clarify the principles upon which they were based before presenting his proposal to the

Lords of the Admiralty. However, after having waited in vain for several weeks for these to arrive, or for further instructions from Michaelis, Best decided that it was now high time to make a formal approach to the Admiralty and duly handed over Mayer's petition, lunar tables, and "Methodus Longitudinum Promota" to the First Sea Lord (Anson) on 20 January 1755.

The additional details required and a wooden model of the repeating circle were duly supplied by Mayer before the end of the year, after which the papers were referred officially by the Admiralty Secretary, Captain Archibald Clevland, to Bradley for his study and comment. A smaller-scale brass model of the repeating circle was made by John Bird according to Bradley's directions and tested by Captain John Campbell, but the instrument appeared to offer no advantage over the Hadley quadrant and never became popular for use at sea. It has, however, been argued that Campbell's invention of the nautical sextant grew out of the experiments which he made with these two types of instruments. For his part, Bradley carefully compared the 230 lunar observations that he himself had made during the previous five years, using Bird's new transit instrument and 8-foot radius astronomical quadrant, with the positions of the Moon predicted for the same times from Mayer's improved lunar tables. He concluded that the latter would probably yield the Moon's longitude to within $\pm 1'$, and therefore strongly recommended them to the Board of Longitude as a means of enabling mariners to obtain quite an accurate estimate of their longitude at sea. Subsequently, he was able to substantiate this opinion by extending the comparison to include over 1100 reliable observations taken at the Royal Observatory with his new instruments.

An autobiographical note by Maskelyne, still in private hands, contains the remark that Bradley's life was

> "probably much shortened by his laborious calculations to verify Mayer's tables by a comparison of them with twelve hundred of his observations, during the five years from the end of 1755 to the end of 1760".[3]

Although this explanation seems to us to be highly unlikely, it is nevertheless more worthy of consideration than the alternative supposition stemming from the French author on Bradley in the *Nouvelle Biographie Générale* that his sudden death on 13 July 1762 was a sign of divine retribution for recommending the adoption of the new-style Gregorian calendar which became legally adopted in England in 1752. Many superstitious people believed that the substitution of the 14th for the 2nd of September 1752, introduced in order to eliminate the error which had accumulated through the previous adherence to the old-style Julian dating, robbed them of eleven days of their lives and blamed the current Astronomer Royal for this imaginary misfortune. As Stephen Rigaud has stressed, the addition of £250 p.a. to Bradley's salary was not intended as recompense for the unpopularity that had thereby been caused, but granted to him in consequence of his having refused in 1751 to take the curacy in the parish of Greenwich since he knew that such an obligation would merely detract from his duties at the Royal Observatory. This had certainly happened in the case of Flamsteed, whose £100 yearly salary had probably been fixed on the grounds that it would later be supplemented by his stipend as rector of Burstow. The additional £250 p.a. awarded to Halley had been a pension granted to him as a retired naval captain, for which Bradley was naturally unable to qualify.

No sooner had Bradley discharged his obligation to the Admiralty by examining and recommending Mayer's tables as a reliable basis for longitude determination at sea, than he became involved in another project—the planning of expeditions to observe the transit of Venus over the Sun's disc in 1761—which was to provide an opportunity for confirming his faith in their predictive ability. The incident which sparked off an interest in observing this rather infrequent event was the presentation to a meeting of the Royal Society on 5 June 1760 of Joseph-Nicolas Delisle's *mappemonde* illustrating the area of the Earth's surface over which the phenomenon would be seen one year later, together with an accompanying memoir on that subject in which Delisle proposed certain changes to the original list of possible observing stations published by Halley in the *Philosophical Transactions* for 1716. Bradley was present at another meeting later that month when the question of how to organize the British expeditions to St Helena and Bencoolen (Batavia) to observe Venus's transit was discussed, and he was asked to list the astronomical instruments required for that purpose. The main items which he proposed were a 2-foot focus reflecting telescope with a Dollond micrometer, a 10-foot focus Dollond refractor, an 8-inch radius quadrant, and a clock or timepiece. These were, however, expensive to purchase, and Bradley was therefore asked to enquire about the terms under which they could be hired. A few days later the Court of Directors of the East India Company promised its full cooperation in placing ships at the disposal of the Royal Society's representatives on the planned voyages and in circulating observing instructions to their representatives abroad. It appeared, however, that there was little or no chance of one of that company's ships reaching the more remote destination in time. Thus after an adverse report on weather conditions at Bencoolen had been received from the local Governor of the East India Company, and Bradley had ascertained that there was no possibility of hiring the instruments required, it was decided to concentrate upon the St Helena expedition.

Bradley's choice of a suitable person to lead this expedition automatically fell on Nevil Maskelyne who, during the previous five years, had devoted many of his leisure hours as a curate in London assisting him in the computation of astronomical refraction. Thus it was Maskelyne who submitted to the President and Council of the Royal Society an estimate of £285 for the instruments to which the expenses of two men's salaries, transportation and subsistence had to be added, bringing the total cost of that expedition to £685. The formal petition addressed to the Lords of the Treasury on 4 July 1760 was for £800, and laid emphasis on the competitive and nationalistic aspects of the undertaking. A similar letter was written by the Earl of Macclesfield to the first Duke of Newcastle (Thomas Pelham-Hollis) one day later; and it was he who took the petition directly to the King. George III immediately issued a warrant for the payment of the expenses for this expedition, and gave a promise to provide the same amount for an expedition to Bencoolen if this should prove to be a practicable proposition.

At Bradley's suggestion the Royal Society resolved that his Greenwich assistant, Charles Mason, should accompany Nevil Maskelyne to St Helena; but after many volunteers had offered to undertake the arduous voyage to Bencoolen, Bradley obtained the Council's permission for Mason to go as principal observer on this latter trip with Jeremiah Dixon as his subordinate. Robert Waddington was appointed as Maskelyne's new assistant. Mason and Dixon were thwarted in their ambition to reach the East Indies, initially because their ship had to return to port after being engaged in a battle with a French man-o'-war in the English Channel, and then because Bencoolen was captured by the French. They were obliged to observe the transit of Venus at the Cape of Good Hope but there they were fortunate in having conditions of perfect visibility. In all of their activities they carefully followed written instructions which Bradley prepared for them. The two observers were able to obtain accurate values for the geographical coordinates of their temporary observing station, using the conventional methods of eclipses of Jupiter's satellites, lunar occultations, and even a lunar eclipse to establish the longitude. Yet they disagreed by several seconds when timing the instants of the internal and external contacts of Venus's disc with that of the Sun, so no new or conclusive information could be drawn from their results.

Maskelyne and Waddington were less fortunate with the weather, and their observations of Venus's transit must on this account be regarded as a total failure. A defective zenith sector was thought to have been responsible for the inconclusive attempt by Maskelyne to determine the parallax of Sirius, which Lacaille's Cape observations had led him to anticipate would be detectable. On the other hand, he made some useful measurements of the rise and fall of the tides in the harbour of St Helena with the assistance of Charles Mason who broke his return journey with Dixon at that island; and the value of the gravitational acceleration relative to that at the Cape was obtained using a seconds pendulum, the results being later standardized by comparison with the transit clock at the Greenwich Observatory. Moreover, during both his outward and return voyages, Maskelyne determined longitude at sea by the method of lunar distances and attained an accuracy of about $\pm 1\frac{1}{2}°$ or 90 nautical miles in one result, as compared with $7°$ in the dead-reckoning procedure of the ship's officer.

The author of the *Dictionary of National Biography* article on Maskelyne gives him credit for introducing this method into navigation; but this is false, since, as we have seen, Lacaille had already practised it during his voyage to the Cape of Good Hope over a decade earlier, and Alexandre-Gui Pingré had been following his example (using various sets of lunar tables) during his own voyage to the island of Rodriguez in the Indian Ocean to observe the transit of Venus. Maskelyne's lunar distances observations were not even the first to have been made using Mayer's tables in conjunction with a Hadley octant, since Carsten Niebuhr, whom Mayer had entrusted with a similar task during a voyage to the Orient, made similar measurements and the appropriate calculations at the beginning of 1760 to obtain the longitudes of Cape Vincent, Cape Spartel, Gibraltar, and Marseilles. These were sent in 1761 from Constantinople (Istanbul) and reached Mayer in Göttingen early in the following year as he lay dying. They impressed him so much that he asked his wife to send them to England after his death, along with a copy of manuscript lunar and solar tables containing further minor improvements that he had made during the course of the previous seven years. This was duly done via the normal diplomatic channels in the summer of the following year.

A Memorial prepared by Michaelis on behalf of Mayer's widow and children, presented by Best to the Admiralty on 1 July and read at a meeting of the Board of Longitude on 9 August 1763, was responsible for the commissioners' resolution to ask two gentlemen well-skilled in astronomy about to leave for the Barbados on H.M.S. *Princess Louisa* to make a second sea-trial of John Harrison's fourth marine timekeeper, to observe eclipses of Jupiter's satellites and lunar distances as frequently as possible during the voyage for the purpose of testing the accuracy of Mayer's new improved manuscript lunar tables. These observations, after being properly attested, would then be considered by the Board and a decision finally taken on the respective merits of the lunar and chronometer methods. The gentlemen in question at that time were Maskelyne himself and Charles Green, previously assistant to Bradley and now to the Rev. Nathaniel Bliss who had been formally appointed as Bradley's successor at Greenwich on 26 August 1762. The new Astronomer Royal was a frequent guest and scientific co-adjutor of the Earl of Macclesfield whose name features, together with those of Martin Folkes, Robert Smith (Professor of Astronomy at Cambridge), James Bradley and William Jones (Professor of Astronomy at Oxford), on an undated testimonial preserved in the British Museum requesting the Archbishop of Canterbury, the Lord High Chancellor, and the other trustees appointed by Sir Henry Saville to nominate Bliss to the Savilian Professorship of Geometry at Oxford.[4] With such influential backing, he had had no difficulty in obtaining that Chair when it became vacant through Halley's death in 1742; nor, with similar support, after Bradley's death some twenty years later,[5] in being elected to replace him at Greenwich. In the interim he had remained in contact with the work being carried out at the Royal Observatory, and assisted Bradley on special occasions such as at the transit of Venus on 6 June 1761 when the latter was prevented by illness from observing this long-awaited phenomenon. His results, that Venus's horizontal parallax is 36·3″ and the Sun's 10⅓″, were published in the second of two letters addressed to the Earl of Macclesfield appearing in the *Philosophical Transactions* for that year.

Aided by the professors of astronomy at Oxford and Cambridge (Thomas Hornsby and Anton Shepherd, respectively), and observed by William Harrison (John Harrison's son), Bliss carefully examined the reflecting telescopes and equal altitude instruments required by Maskelyne and Green, and judged them to be sufficient for the purpose of finding longitude at sea, or the time at a land-based observing station. Maskelyne was regarded as a trustworthy exponent of the method which Mayer had been championing, especially since, shortly before he began his voyage to the West Indies on 15 November 1763, his *The British Mariner's Guide*, containing full details of the theory and practice of the lunar distances technique appeared in print. In this book, to which Mayer's *first* set of lunar tables are appended, he proposed that the British should adopt Lacaille's plan for a nautical almanac—the first indication of the important development which was yet to come.

After returning from the West Indies in the autumn of 1764, Maskelyne discovered that Bliss had just died and that he had already been strongly recommended by the Earl of Morton (now President of the Royal Society) and other influential friends in London and Cambridge as the most suitable successor. He returned Mayer's tables to Bliss's widow who sent them to the Chairman of the Board of Longitude (Lord Egmont). Egmont, in turn, passed them on to John Bevis and George Witchell who required them for calculating lunar distances. Longitude differences obtained by these and another two computers, Captain John Campbell and William Harrison's nominee James Short, were then obtained from a comparison of Maskelyne's and Green's observations at Bridgetown of nine emersions of Jupiter's first satellite with five emersions of that same satellite observed at Portsmouth by John Bradley (the late James Bradley's nephew and faithful assistant for fourteen years), and two observed by Bliss at Greenwich. Those differences were also deduced independently from the declared clock-rate of Harrison's timekeeper and equal altitude observations made at Portsmouth and Bridgetown, and the results recorded in the following form in the minutes of a meeting of the Board of Longitude on 19 January 1765:[6]

Difference of Meridians between Portsmouth and Barbadoes

Computers' Names	by Observation			by Timekeeper			Error of Timekeeper		
	H	′	″	H	′	″	H	′	″
Captain Campbell	3	54	19·1	3	55	00·6	0	00	41·5
Dr. Bevis	3	54	22⅛	3	54	56	0	00	33⅔
Mr Witchell	3	54	12·7	3	55	00·05	0	00	47·3
Mr Short	3	54	18½	3	54	53	0	00	34½

These figures were corroborated by further independent calculations by five other well-qualified members of the Board, and the commissioners unanimously agreed at their next meeting three weeks later that Harrison's watch had indeed kept time to well within the strictest limit imposed in the 1714 Act. The mean error of $+39^s·2$ was equivalent to a longitude difference of only 9·8 nautical miles, whereas the greatest discrepancy shown in Witchell's values was only 11·8 nautical miles. Thus Harrison's pride in his masterpiece was fully justified.

Maskelyne submitted the results of the lunar distances computations in the form of a petition to the Board of Longitude at a meeting on 9 February 1765, where he arranged for four ships' officers to be in attendance to testify verbally to their successful application of the instructions given by him in *The British Mariner's Guide*. He began by declaring that he had made frequent observations of those distances during his previous voyage to and from St Helena which, in conjunction with Mayer's original set of lunar tables, had enabled him to compute longitude consistently with an accuracy of less than 1° or 60 nautical miles; and that he had already explained his procedure fully in his book. In his voyage to the Barbados and back, he had made similar observations which, after being carefully corrected for instrumental errors, yielded the longitude at sea to within ½° or 30 nautical miles. The only difficulty was the complexity of the calculations, which he himself had taken up to four hours to complete. He therefore suggested that if Mayer's "last manuscript Lunar Tables" were to be printed in a nautical ephemeris and auxiliary tables provided to correct the measurements for refraction and parallax, the longitude at sea could certainly be found to within the wider of these two limits. The Board approved this proposal, and resolved to apply to Parliament for a sum not exceeding £5000 for Mayer's widow as well as £10,000 for Harrison.

This financial assessment of the relative value of the lunar distance and chronometer methods would appear to reflect the Board's recognition of the superior accuracy attainable by the latter; also its refusal to acknowledge that either was generally useful until such a publication as Maskelyne had mooted came into existence, or until Harrison had disclosed the principles of his watch's construction. Moreover, two copies of the latter's masterpiece capable of performing with a comparable degree of accuracy had to be constructed by other craftsmen and then tested at the Royal Observatory. It was certainly Clairaut's protest in *The Gentleman's Magazine* for May 1765, to the effect that both he and Euler deserved similar recognition to Mayer for their fundamental theoretical contributions to the development of the lunar theory, which caused Parliament to reduce the amount paid to Mayer's widow to £3000 and award an unsolicited £300 to Euler. There may, however, have been a misunderstanding regarding the precise nature of Mayer's debt to Euler since the former's basic equations were borrowed not from the latter's lunar theory but from his prize essay to the Paris Academy in 1748 on the inequalities in the motions of Jupiter and Saturn. The Board's decision at this memorable meeting to pay the second half of the maximum sum to Harrison only after he had supplied written explanations and diagrams of his watch's construction, and revealed its principles to the satisfaction of three gentlemen skilled in mechanics and three watchmakers, was perfectly reasonable, but it was destined to spark off a violent personal controversy between the ageing inventor and Maskelyne which was partially responsible for the former being accorded the name 'Longitude' Harrison.

By this time, Maskelyne's plan for the nautical ephemeris had been duly approved. Four computers, Israel Lyons, George Witchell, John Mapson, and William Wales, were accordingly employed under his direction to calculate the required ephemerides for the years 1767 and 1768. Each was to receive £70 on the satisfactory discharge of this work, in instalments to be meted out from time to time in proportion to the progress being made; and they were also to be furnished with such books and tables as would be necessary to assist their calculations. Richard Dunthorne of Cambridge, whose early published work on the lunar theory had been used by Mayer as an additional vindication of the accuracy of his own tables, was appointed as the comparer of the ephemerides and corrector of the press. It was his duty to examine the individual computations which were transmitted to him monthly by the Astronomer Royal, and to give the computers such instructions as they might need to execute their work in conformity with the approved plan. He was also responsible for the selection of suitable zodiacal stars from which the Moon's distance at three-hourly intervals was to be calculated; and for forecasting the eclipses of the Sun and Moon, and occultations of stars by the Moon, that would be observable at Greenwich. Moreover, he was to "make the configurations of Jupiter's satellites and the other phenomena of the ephemeris" and to include part of this additional information in the almanac. The astronomical ephemerides which, together with other relevant tables and explanations, constitute the *Nautical Almanac*, were first published in December 1766 and a thousand copies immediately circulated in the United Kingdom, North America and Europe.

Meanwhile, Maskelyne had been editing Mayer's manuscript "Theoria motuum lunae" (transmitted to England late in 1755) containing the theoretical framework in which the manuscript solar and lunar tables had been constructed. In addition he made numerous amendments to these tables themselves, such as reducing Mayer's calculations from the Paris to the Greenwich meridian, and adopting his figure of $9^m 16^s$ for the longitude difference. Moreover, he recalculated the tables of the Moon's hourly motion in longitude, and converted the thermometric scale in the general algebraic formulae and tables of astronomical refraction from Réaumur to Fahrenheit degrees. When considering the variations in the effect of refraction due to atmospheric pressure, Mayer had used an arbitrary but uniform barometric scale in which one division caused a variation in refraction equal to two-thirds of that corresponding to an interval of $1°$ Réaumur. Maskelyne recalculated these variations in terms of the absolute barometric scale in inches of mercury. He also gave appropriate titles to tables for computing the equatorial coordinates of the Moon and planets and to the lunar tables themselves. Two thousand copies of the amended tables were then printed under his personal supervision, and stored at the Royal Observatory until 1770, when they were published in both Latin and English at a cost of ten shillings each.

Barely had Maskelyne succeeded in getting the first issue of the *Nautical Almanac*, Mayer's *Theoria Motuum Lunae* and the *New Tables of the Motions of the Sun and Moon* all into print before the end of 1767, than he became involved in the preparations for the forthcoming transit of Venus due to occur two years later. At a meeting of a special Royal Society Committee on 17 November 1767 he was among four Fellows who read papers on this subject, the others being John Bevis, James Short and James Ferguson; and it was he who presented a final report two days later embodying recommendations based upon their respective statements. These were to the effect that British observers should be sent to Fort Churchill on the western side of Hudson's Bay, to the small island of Vardö off the northern coast of Finland, and to the northernmost tip of the Scandinavian peninsula. Two more observers should go to an island in the South Seas somewhere within the area specified by Thomas Hornsby in the *Philosophical Transactions* of 1765. Each of these four teams should be equipped with a quadrant, a clock, two 2-foot reflecting telescopes with micrometer attachments, a thermometer, a barometer, and a compass. The best men qualified to undertake such a task were John Bradley, Charles Mason, Jeremiah Dixon, Samuel Dunn, Charles Green, Joseph Dymond, William Wales and Alexander Dalrymple. Dymond and Wales were selected for the Hudson's Bay station, and their temporary observatory—an octagonal construction with brass rings and rollers for a slotted roof —was designed by Maskelyne with the cooperation of John Smeaton and the Greenwich carpenter, Mr Ashworth. The Hudson's Bay Company promised to provide the transportation.

After the King had approved the Royal Society's plans, and honoured its request for £4000 for the entire undertaking, the Admiralty initiated its own activities in connection with the South Seas programme. James Cook was appointed Captain of the bark *Endeavour* and Charles Green was elected to go on that expedition. Joseph Banks and Karl Solander accompanied them in order to further their interests in natural history, and the island of Tahiti was eventually chosen as their destination. Maskelyne proposed that his assistant William Bayly ought to be considered, and Bayly travelled north with Dixon before they parted company to observe at the North Cape and the island of Hammerfest respectively; and it was also at his request that Mason went to Cavan, near Strabane in Donegal (north-west Ireland) to observe the transit. He himself observed the ingress of Venus at the instants of outer and inner contact with his 2-foot reflector at Greenwich. These observations represent only a minor fraction of the total 150 made by British,

Danish, Dutch, French, German, Russian, Spanish, and Swedish astronomers; and the papers published on this subject in the *Philosophical Transactions* constitute an even smaller fraction of those which appeared in the Paris Academy's *Mémoires* and in other scientific periodicals. The extreme values deduced for the solar parallax on this occasion were the Swede Anders Planman's 8·43″ and Alexandre-Gui Pingré's 8·80″ which contained a much narrower range than the limits of 8·28″ and 10·60″ found for the previous transit. The higher degree of internal consistency among the results of the 1769 transit may be ascribed partially to the general improvements being made in the manufacture of instruments; in particular, to the use of achromatic telescopes. The achromatic lenses patented by John and Peter Dollond brought even greater precision to positional astronomy, not only because they removed colour aberration, but also because they gave sharp and bright images of the stars over a larger field of view than did contemporary reflecting telescopes. Astronomers were consequently encouraged to provide telescopes of a given length with a larger aperture and thereby to extend their observations to fainter stars.

The most serious problem affecting the measurements made of Venus's position during its transit was not instrumental, but rather physiological in character. Owing to the great contrast in the intensity of the light from those two heavenly bodies, the human eye sees Venus in a distorted form at the crucial instants when it enters and leaves the solar disc, giving rise to an optical aberration which is commonly referred to as 'the black drop'. This phenomenon could produce errors of up to ±1 minute in observations that were expected to be correct to within ±1 second. Another kind of uncertainty was also incurred in the manner of reducing the widely differing values resulting from this troublesome appearance. Anders J. Lexell used a new analytical method by Leonhard Euler to evaluate the results of Green's, Cook's, and Solander's observations off Tahiti, and arrived at the value of 8·63″. Later, however, Simon Newcombe, aided by his knowledge of more accurate longitude determinations, revised this figure to 8·79″ which is less than one-hundredth of a second of arc from the modern accepted value. Other sources of inaccuracy were, of course, inherent in the design and construction of the telescopes employed, which caused the proven skills of eminent mathematical instrument-makers such as George Graham, John Bird, and Jesse Ramsden to be in very great demand. Indeed, the last two between them were to supply most of the needs of both British and foreign astronomers during the latter half of the eighteenth century.

In 1767, Bird disclosed to the Board of Longitude his method of graduating and dividing the brass mural quadrant installed seventeen years previously at the Royal Observatory, and 500 copies of his written description were printed in the following year. After having received an order from Bradley in February 1749 to construct this instrument, Bird had examined the iron quadrant made by Graham and concluded that an error of 16″ over the whole 90° arc had been caused by the manner in which the different pieces had been fitted together. Brass, he hoped, would be less susceptible to deformation under the quadrant's own weight; and after having tested it with Bradley on a hot summer's day just over ten years later, he felt able to claim that "... neither hot nor cold weather, nor the weight of the instrument, have any material effect upon its figure".[7] He expected therefore that during the time which had elapsed since it was first set up and the time of his writing, there had been no sensible alteration to its shape; the errors of the divisions and in reading off the answer being no greater than ±3″. Bird followed Graham's principle of bisecting but not trisecting or quinquesecting linear or angular distances. His mental attitude towards his work may be inferred from the following quotation from his preface to *The Method of Dividing Astronomical Instruments* (London, 1767):

> "How far the Lunar Theory hath been improved by the observations of the late Dr Bradley, and Mr Mayer, I leave to the decision of those who have tried it by observations, in order to find the longitude at sea, &c. I cannot help, however, being fully of (the) opinion, that a still more perfect knowledge of the motion of the heavenly bodies may be obtained by future observations, skilfully made, with accurate Instruments."[8]

As regards nautical instruments of smaller radius than the large quadrant made by Bird for the Royal Observatory, the process of dividing was at last becoming mechanized. Ramsden's 'Engine', described by him with a preface signed by Maskelyne on 28 November 1776, had been designed for that very purpose.

It consisted of a large wheel of bell-metal supported on a mahogany stand, having three legs braced together. On each leg was a conical friction-pulley on which the dividing-wheel rested.

To prevent the wheel from sliding off these pulleys, the bell-metal centre under it turned in a socket on the top of the stand. The circumference of the wheel was cut into 2160 teeth, meshed with an endless screw. Six revolutions of the screw therefore turned the wheel through $(6 \times 360°)/2160 = 1°$. A brass circle fixed to the screw arbor had its circumference divided into 60 parts, so that each division corresponded to an angular displacement of $10''$. Different arbors made of tempered steel were available when different degrees of accuracy were required.

The instrument to be divided could be screwed to the dividing-wheel through holes made in the latter's radius; the centre of the former being fitted exactly on one of the arbors, with the frames carrying the dividing-point attached to that carrying the endless screw at whatever distance from the centre of the wheel that the radius of the instrument to be divided may require, then clamped in that position. This method was intended to prevent errors from the expansion or contraction of the metal during the dividing process. The usefulness of this invention encouraged Ramsden to evolve a similar method for dividing lines into uniform scales to within 1/4000 inch, or in the ratios of circular functions which he described in a second pamphlet printed two years later. The critical component of this 'Engine' was the accuracy with which the endless screw could be cut, since its whole length was engaged in the teeth of the movable plate and not just a few threads at a time as in the case of the circular engine. The fact that these operations could now be performed by anyone—not necessarily by an expert—was important in increasing the numbers of reliable nautical instruments which could now be constructed in a given time. A consequence of this uniformly high degree of reliability was that the radius of a Hadley quadrant or sextant could now be halved, and thereby rendered more manageable, without any deterioration in accuracy— a great boon to the practising navigator, and an important step towards making the method of lunar distances generally useful as well as practicable as required by the 1714 Act.

F

NOTES

1. Delambre, 1827, p. 421.
2. Forbes, 1971 a, p. 79; quoting from Euler to Mayer, 26 February 1754.
3. N.M.M. MS.: PGR/38/1.
4. B.M. Add. MSS. 35,587, f. 18^r.
5. B.M. Add. MSS. 38,199, f. 42^r. This is Bliss's letter of application, dated Oxford, 29 July 1762, which was probably addressed to the first Earl of Liverpool.
6. R.G.O. MSS. 533, Board of Longitude Confirmed Minutes, 5 (19 January 1765), p. 35.
7. Bird, 1768, p. 24.
8. Bird, 1767, pp. v–vi; quoted from Hellman, 1932.

7

An Age of Improvement

WE have seen in the previous chapter how Maskelyne, at the beginning of his career as fifth Astronomer Royal, devoted most of his energies to instituting a nautical alamanac, in accordance with a proposal made over a decade earlier by the Abbé de Lacaille. The main feature of this annual publication was the table of lunar distances computed for three-hourly intervals, which served as a basis for the determination of longitude at sea. But he regarded it as a great discredit and dishonour to the British nation, the Royal Society, and the general state of learning in Great Britain, that Lacaille's star catalogue and Mayer's lunar tables had to be used as necessary prerequisites for the calculations. Why should foreigners have been able to succeed in what the previous Astronomers Royal at Greenwich, equipped with far superior instruments, ought to have achieved? A brief indication has already been given in Chapter 6 of the positive reasons for Lacaille's and Mayer's important contributions.

An explanation of why British astronomers failed to keep abreast of these continental rivals is provided by Maskelyne himself in a memorial read to the Royal Society at its meeting on 9 June 1763, during Bliss's term of office as Astronomer Royal and between the times of his own expeditions to St Helena and Barbados. In it he attributes the blame largely to the attitude of Bliss's three illustrious predecessors towards their observations being their own exclusive property. Flamsteed's tenaciousness in withholding his observations not only from the public but also from his learned friends was responsible, said Maskelyne, for a delay in Newton's establishment of his lunar theory; the situation being remedied only by the actions of the Royal Society described in Chapter 3. Had Flamsteed's data been published regularly (and subjected to regular scrutiny and criticism), instrumental errors might have had less effect on their accuracy. Unless the Royal Society had acted in retrieving Halley's observations from his executors, the whole series of 19 years' lunar observations might also have been lost. Maskelyne goes on to claim that had British astronomers received the same lunar observations which had been communicated from Bradley (via Euler and Schumacher) to Mayer, perhaps one of the former would have been equally successful in perfecting the lunar theory on Newtonian principles—a dubious supposition, in view of the retarded state of analytic methods in Britain at that time.

Bradley's denial of his observations on Sirius to Maskelyne before the St Helena expedition meant that the latter had to rely entirely upon Lacaille's Cape observations for 1751 and 1752 published in the latter's *Fundamenta Astronomiae* (Paris, 1758) which seemed to favour a parallax for that star of about 8″. Now that Bliss had provided him with an extract from Bradley's data which revealed that no appreciable parallax had been detected, he realized that he had been wrong in placing so much faith in the accuracy of Lacaille's zenith distance measurements. The Royal Society had thereby incurred unnecessary expense in purchasing a 10-foot zenith sector for his investigation of the hypothetical effect, and he himself had wasted much time and effort in searching for it. Thus, as a consequence of such jealous guarding of the Greenwich observations, the progress of astronomy in Britain was being retarded, the principal object of the Royal Observatory's existence was being defeated, and foreign scientists were beginning to get a distorted view of British scientists —thinking them to be

> "capable only of abstruse Theories, but either not able, or not industrious enough, to make assiduous & repeated experiments & Observations, tho' the only true Basis of Natural Philosophy".[1]

The remedy to this situation lay in the hands of those responsible for the governing of that astronomical institution; namely, the Royal Society itself in the first instance and, through their Board of Visitors, the Lords of the Admiralty, the Board of Ordnance, and ultimately King George III himself. Yet despite Maskelyne's timely and emphatic appeal for the *absolute necessity* of immediately demanding Bradley's observations from his executors, their representative William Dallaway was able to visit Greenwich fifteen months later, after Bliss's death, and take Bradley's observation books away; with what effects, we have already seen in Chapter 5.

The tragic deed which Maskelyne had striven to avert was thus enacted before his return from the West Indies, and was to remain the source of bitter disappointment and constant irritation to him throughout the remainder of his career. On the other hand, as a result of his initiative, steps had begun to be taken in his absence to prevent a recurrence of such crippling blows to the efficient working of the Royal Observatory. By a small majority the Royal Society had resolved to finance the publication of previous observations whenever these should become available and, this time by an over-whelming majority, to compel the existing and future Astronomers Royal to honour the obligation imposed by Queen Anne's Warrant of 12 December 1710 to submit each year's observations within the first six months of the following year. Newton, less than three weeks before his death on 20 March 1727, had publicly reprimanded Halley for not doing so, but this would appear to have been the one and only occasion of such a demand being formally made since 1714, in Flamsteed's time.

By now, of course, it was recognized that what was required was a more detailed set of regulations clearly defining the responsibilities of the Astronomer Royal's office. The animosities of the earlier years between Flamsteed and several Fellows of the Royal Society (cf. Chapter 3), the frustrations experienced by Halley on taking over the directorship of the Royal Observatory in 1720 (cf. Chapter 4), and the legal complications associated with the recovery of Bradley's observations (cf. Chapter 5), might all have been averted if such rules had been prescribed at the beginning. It was no excuse to maintain that the Observatory had managed to function for almost ninety years without them, since it had done so very imperfectly. Bliss's death, followed by the confiscation of Bradley's observations, undoubtedly acted as catalysts in the Royal Society Council's drafting out of the following nine regulations for His Majesty's consideration on 8 November 1764:[2]

1. The Astronomer Royal must reside at Greenwich, and not take up any outside duty. Any absence from his post must be approved by the Board of Visitors.
2. The Astronomer Royal and his assistant must never be absent from the Royal Observatory at the same time.
3. No one is to be allowed in the instrument room unless the Astronomer Royal or his assistant is present, to avoid the possibility of damage.

4. Henceforth, no money would be taken from those who were admitted.
5. The Astronomer Royal had to record faithfully all his observations in his observation books, and put his initials at the foot of the page.
6. The Assistant Observer had likewise to keep a similar record of his own observations, and authenticate them in the same way.
7. The observation books had never to be removed from the observatory.
8. A true and fair copy of the annual observations was to be sent to the Royal Society Council within six months after every year.
9. The Visitors' orders to the Astronomer Royal and his assistant were to be obeyed.

There were obvious reasons for insisting upon the Observatory never being left vacant; for besides reducing the security risk it ensured continuity in the observations which it was the prime duty of the Astronomer Royal and his assistant to collect. Forbidding entry of unauthorized persons to the 'instrument room'—presumably the quadrant room, where the 8-foot diameter transit circle and the two 8-foot radius quadrants made by Graham and Bird were housed—was again a useful precautionary measure; but forbidding the assistant to augment his meagre salary by showing round visitors, as Bliss's assistant Gael Morris had been permitted to do, was regarded by Maskelyne as an unnecessary imposition. Indeed, William Bayly, who replaced Maskelyne's first assistant Joseph Dymond when the latter left him in November 1766, threatened to resign less than five years later because he could command over three times as high a salary as a schoolmaster at the Royal Academy in Portsmouth, and was given no opportunity of earning as much at Greenwich. He was, however, to find a more lucrative and adventurous employment as astronomer on Captain Cook's second voyage.

The importance of keeping a careful record of who made which observations was subsequently to be impressed upon Maskelyne when he discovered that an assistant named David Kinnebrook had begun to record the times of transit between $0^s{\cdot}5$ and $0^s{\cdot}8$ later than his own observations. This discrepancy was detected through the daily rate of the transit clock, deduced from a star observed on two days by each of them, being different from that based on another star observed by only one of them. Bradley's method of noting the proportional distance of a star from a horary wire at the two beats of the clock immediately preceding and following the star's passage across the wire was used, and the transit time estimated to the nearest $0^s\!\cdot\!1$. A systematic time-lag of several times this magnitude was regarded by Maskelyne as intolerable, *whatever the cause*, thus he reluctantly decided to dispense with Kinnebrook's services, although he had no other reason to be displeased with him.

The regulation affecting the preservation of the observations and enabling other qualified persons to have access to them was later invoked by the Royal Society's Assistant Secretary, Stephen Lee, who made use of this privilege to provide an embarrassing critique of the Royal Observatory's work which only the force of authority prevented him from publishing. This affair will be discussed briefly in Chapter 8. The penultimate regulation was the one which Maskelyne himself had pressed to a vote two years previously, and the last had to be obeyed unless the warrant of the late Queen Anne and the powers given by it to the Royal Society's current Board of Visitors were to be flaunted.

Only two days after his official appointment as Astronomer Royal through the Royal Warrant of 26 February 1765, Maskelyne gave a written assurance to the President and Council of the Royal Society that he would now begin to support by deeds the views which he had expressed less than two years earlier. While indicating his approval of the draft resolutions, he expressed the hope that no others would be imposed upon him which might incur a weakening of respect on the part of his assistant and check the exertion of his own zeal and endeavours to improve the state of astronomical science. This rather vague statement may be interpreted as an indication of his dissatisfaction with the fourth regulation affecting his assistant's salary; but a more likely explanation is that he had already received prior notice of an intention to add a tenth regulation, duly interposed between those numbered 2 and 3 on the draft copy in the final version authorized by George III on 12 March 1765, that he was personally and directly responsible for ensuring the diligence of his servant or assistant.

Four days later, in response to Maskelyne's own request, the Board of Visitors (including Captain John Campbell and James Short) came to the Royal Observatory to prepare an inventory of the mathematical and astronomical instruments there, the previous one having been compiled on 18 August 1762 shortly after Bliss had taken up office. A second survey was carried out soon afterwards by Maskelyne with the aid of Bird and Short, and their estimate for repairs to the instruments and buildings duly submitted to the Board of Ordnance which reluctantly consented to accept the responsibility of payment. That same year (1765), a large collection of Halley's manuscripts, including four of his five observation books (cf. Chapter 4), were presented to the Royal Society by his daughter Mrs Catherine Price, and deposited at Greenwich; for which gesture she was awarded £100 by a grateful Board of Longitude. Just over six years later Mrs Elizabeth Tew of Islington who, as the widow of James Hodgson's representative, had inherited the vast quantity of Flamsteed's papers, received the same compensation from the same source. Bliss's observations were obtained by the Royal Society from his widow, to whom 50 guineas were paid from its own funds on account of her straitened circumstances. Meanwhile, the lawsuit for the recovery of Bradley's observations, the most important of all, still lingered on—despite Maskelyne's continual complaints and personal involvement in the Royal Society's attempts to accelerate publication (cf. Chapter 5).

But although Maskelyne did not possess Bradley's data, he had fallen heir to his equipment and, through his former association with Bliss, knew a great deal about Bradley's observational techniques. Most of his time was spent using the 8-foot transit instrument which Bird had fitted with a single eye-glass of 50 times magnification. Five wires, as exactly parallel to one another as that skilled instrument-maker could manage to arrange them, were placed in the common focus of the telescope; the middle or meridian wire being made to pass through the centre of the field of view. A sixth equatorial wire—or directing wire, to use Maskelyne's own terminology—crossed at right angles and adjustments were made until the star for which the transit time was required appeared to drift along it as a result of the Earth's diurnal rotation. The four additional horary wires were used either to check or to supplement the observation made on the meridian wire; the mean of the times of transit of a star over two wires at equal distances from, and on different sides of, the meridian wire being the same as the time of meridian transit. A further use of these additional wires was that they served to multiply the results for the clock-rate found by the difference in transit times of the same stars over the several wires on different days. The time of a star's or planet's transit could, exceptionally, be inferred from a single observation of its appulse to any one of the wires, since the difference in time corresponding to the spacing between the wires was known from taking the mean of a great number of observations.

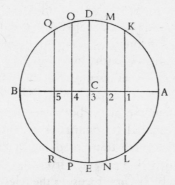

Fig. 6

The above arrangement is illustrated in Fig. 6, in which ADBE represents the cell containing the wires stretched in the focus of the eye-glass. MN, OP are the two nearest and KL, QR, the two farthest wires on either side of the meridian wire DE representing portions of horary circles. The directing wire AB, which we suppose provisionally to be exactly at right angles to each one of these three wires, will consequently lie in a parallel of declination: it is the line along which a star was made to drift by suitably adjusting the telescope and rotating the cell. Since A denoted the eastern side of the field of view, B the western side, D the southern side and E the northern side, the observed motion is along AB (from right to left). Maskelyne adopted the convention of numbering the horary wires consecutively in the same direction; that is, the five wires KL, MN, DE, OP, QR were designated by him as wires 1, 2, 3, 4, 5 respectively. The transit times of a star across each of these successively were noted to the nearest $\frac{1}{4}^s$—an accuracy which was later improved by introducing finer wires—and this observation was repeated on different nights until a reliable set of average values for the four respective sidereal time differences 1–2, 2–3, 3–4, 4–5 could be obtained. The corresponding mean differences in right ascension were then found by dividing those values by the cosine of the star's declination.

The existence of optical aberrations in the double eyepiece which he initially used caused Maskelyne to place greater weight upon the inner wires (2 and 4) than upon the outer ones (1 and 5); but after Bird in August 1773 had modified the nearest eye-glass to slide in its cell, the celestial object and the cross-wires could both be brought into sharp focus over the entire field of view and differences in times of all five wires averaged equally. Any slight lack of parallelism between the wires was found by comparing four mean differences for stars near the Equator with those found for others of higher declination, making due allowance for the convergence of the horary circles towards the celestial pole. The effect of any small deviation from 90° in the angle between the directing wire and middle horary wire could be estimated by rotating the eyepiece until these two wires were transposed, and taking half the mean difference in right ascension obtained from the two sets of observations; or, alternatively, by first focusing the telescope upon a distant terrestrial object such as a carefully drawn cross with its arms truly at right angles to one another to find the angle of deviation, then calculating the correction to be applied in right ascension from spherical trigonometry. An increase in magnification to 80 times was achieved through Maskelyne's replacement of the original object-glass by an achromatic doublet of 2·7 inches aperture. By also yielding brighter and more distinct star images this new lens enabled the transit times to be measured with an even greater degree of precision. As has already been remarked, the method employed by Maskelyne for estimating fractions of a second in his timing of transit observations was to judge the proportion of the distances of a star from one of the horary wires at the beats of his clock immediately before and after the instant of transit. Until 13 September 1772, he estimated these fractional parts to the nearest $\pm\frac{1}{8}^s$, but then began adopting a decimal sub-division as being more convenient for calculation. He had confidence in his timing to about $\pm0^s\cdot2$; mistakes amounting to 5, 10, or even 30 seconds usually being detectable by repeated measurements.

Other miscellaneous observations recorded in Maskelyne's transit book included eclipses of the Sun and Moon, eclipses of Jupiter's satellites, and occultations of stars and planets by the Moon. These were apparently made with one of the instruments in the Great Room, which included a 6-foot focus 9·4-inch aperture Newtonian telescope, a 2-foot focus 4·36-inch aperture reflector by Short, and a 46-inch focus 3·6-inch aperture achromatic telescope with a triple object-glass and divided achromatic object-glass micrometer by Dollond. This was a new contrivance invented by Servington Savary, applied by John Dollond (senior) to the object end of a *reflecting* telescope, and by Peter Dollond to an achromatic refractor.

Most of Maskelyne's observations of the eclipses of Jupiter's satellites were made with the last-mentioned instrument, because of the excellence of its optics and the advantage of its micrometer over the conventional type for measuring small angular distances. The common wire micrometer was adapted primarily for measuring differences in right ascension and declination. Bradley's own manuscript directions on how to set such an instrument on a star and calibrate it for use in telescopes of different focal lengths were read by Maskelyne to the Royal Society on 20 February 1772 and afterwards published in the *Philosophical Transactions*. Two months previously, he had also described how the object-glass micrometer could be applied to the measurement of small differences of right ascension and declination between the limbs of the Sun and Venus or Mercury, and to observing the distances of the limbs in lines both parallel and perpendicular to the Equator during the transits of these planets over the Sun's disc. A copy of the seven instructions for using Dollond's micrometer which had been issued to the Royal Society observers who had viewed the 1769 transit of Venus at the North Cape of Scandinavia and in the South Seas, as well as to those who had gone to Hudson's Bay and the North of Ireland, was appended to this explanation when it was published in the same periodical for the year 1771. A few years afterwards Maskelyne, aided by Peter Dollond, investigated the effect of placing prisms between the object-glass and eye-glass of a 30-inch long achromatic telescope, and invented the prismatic micrometer (cf. Fig. 13), which superseded the use both of the vernier scale and common micrometer. It was a device of which he was justly proud, since it proved capable of measuring horizontal refractions, the dip of the horizon, and small altitudes and depressions of land objects, with an even greater accuracy than had ever before been attained.

Maskelyne's comparison of the eclipses of Jupiter's first satellite obtained with the above-mentioned Dollond refractor and two reflecting instruments revealed that the immersions occurred soonest with the Newtonian, 13^s later with the Dollond, and 7^s later still with the Short telescope; converse results were found for the emersions. This source of systematic errors was undoubtedly largely to blame for a significant disagreement between two independent determinations of the longitude difference between Paris and Greenwich using the same method reported by Peter Wargentin in a letter to Maskelyne written at Stockholm on 19 March 1776, which was read to the Royal Society almost eleven months later and published in the *Philosophical Transactions* for 1777. The mean of seven immersions and eight emersions observed by Charles Messier in Paris with either a 2½-foot Gregorian reflector or 3½-foot achromatic refractor, and by Maskelyne at Greenwich using the 6-foot Newtonian telescope, gave 9^m 38^s as their separation in longitude, although other averaging could produce appreciable changes in this figure. Wargentin, using a 10-foot Dollond achromatic of 90 times magnification to observe eight corresponding events between 1761 and 1764, had found the Stockholm–Greenwich longitude difference to be 1^h 12^m 21^s. From solar eclipse observations by himelf and the Paris astronomers Gui Pingré, Dionis Du Séjour, and Anders Lexell, he knew the Stockholm–Paris longitude difference was 1^h 2^m 55^s; thus he inferred that the Paris–Greenwich separation was 9^m 26^s, a discrepancy of 12^s, which was indicative of the limited reliability of this phenomenon as a basis for longitude determination.

The times of these eclipses were noted by Maskelyne from a clock in the Great Room, but afterwards reduced to those on the transit clock. Both of these clocks had been constructed by John Shelton: they were generally wound up on the first day of each month, and regulated to show sidereal time, since this was convenient both for stellar observations or for deducing the difference of right ascension between different stars, or between stars and planets, from the recorded time differences. The 'carrying' of the time of one to that of the other for the purpose of intercomparison was not done by means of a pocket-watch but merely by opening the windows wide, and listening in the transit room to the sound of a bell on a so-called 'assistant' clock in the octagonal room which rang every time the second hand reached 60! A good ear was therefore just as important as a good eye in this kind of observation, or indeed even when listening to the ticking of the transit clock while taking meridian observations.

Every observation of the meridian transit of a planet was usually accompanied by another of its meridian zenith distance, made with Bird's 8-foot radius brass mural quadrant. These observations, together with others made with Graham's 8-foot radius iron quadrant of stars culminating north of the zenith, were noted in a separate book from that in which the transits were set down. Each of these quadrants was divided into 90° by the interior divisions, and by 96 parts by the exterior divisions; the latter number being a consequence of Graham's method of continual bisection (cf. Chapter 6). Each degree was subdivided into 12 parts or to every 5', then into every ½' or 30" by a vernier attachment, and finally by a micrometer screw into every 1". Each of the 96 parts was divided into 16 parts, then each of these into 16 smaller parts by another vernier, and finally by a micrometer screw to every second. Two convex lenses placed exactly opposite the two verniers magnified both sets of scale divisions, thereby rendering them easier to read. Maskelyne provided a table for facilitating the conversion of this exterior scale into degrees, minutes, and seconds, so that a quick reduction could be made and used as a check upon the reading obtained from the interior scale. He invariably adopted the exterior scale as the more reliable of the two, recording in every case the height of the barometer and two Fahrenheit thermometers—one indoors and the other outside but in the shade—to enable him to compute the refractions.

The theoretical basis for these calculations had been taken by Maskelyne from one of Bradley's manuscripts containing calculations for the latitude of the Greenwich observatory and a formula suitable for logarithmic work derived from a rule stated in Thomas Simpson's *Mathematical Dissertations* (London, 1743). Simpson's derivation of this rule invokes the incorrect assumption that the density gradient in the Earth's atmosphere decreases uniformly with altitude above sea-level; yet the proposed formula which, in common with others used by Lacaille and Mayer, takes account of the barometric and thermometric corrections, was found in practice to yield such reliable predictions that it soon became widely adopted, and remained in use until well into the nineteenth century.

Bradley's rule was first published, though without proof, by Maskelyne in *The British Mariner's Guide* (London, 1763), and demonstrated by him one year later in the *Philosophical Transactions* for the benefit of astronomers and the few mariners then capable of understanding it. The rule was repeated in Maskelyne's "Requisite Tables" (1767), in the *Nautical Almanac* for 1767 where a table of mean refractions based on it is also given, and again in the *Astronomical Observations made at the Royal Observatory at Greenwich from the Year 1765 to the Year 1774* (London, 1776), the first of four such volumes published during his lifetime. The statement of it in his "Explanation and Use of the Tables" is as follows:

> "the refraction at any altitude is to 57 seconds, in the direct compound ratio of the tangent of the apparent zenith distance, lessened by 3 times the refraction, to the radius, and of the altitude of the barometer in inches, to 29·6 inches; and in the reciprocal ratio of the height of Fahrenheit's thermometer increased by the number 350 to the number 400."[3]

The figure of 57″ for the mean refraction of 45° 3′ had been inferred from calculations in which the Sun's horizontal parallax had been taken as $10\frac{1}{3}$″ instead of 8·8″ as found from the two transits of Venus. A substitution of the latter value (which coincides with our modern one) would reduce that constant by $\frac{1}{2}$″ to $56\frac{1}{2}$″ and increase Bradley's value for the latitude of the Royal Observatory by the same amount to 51° 28′ 40″.

A source of error implicit in the Greenwich lunar observations for which Bradley had made no correction, was the change in zenith distance liable to occur in a matter of seconds owing to the inclination of the Moon's orbit to the celestial equator (or parallel of declination). Maskelyne saw the necessity of either noting the zenith distance of the Moon's upper or lower limb at the instant when the illuminated easterly or westerly limb crossed the meridian, or taking account of the time-interval between observations made with the two instruments in order that the difference in declination could be computed and the observed zenith distance reduced to the time of transit of the easterly or westerly limb. His observations with Bradley's $12\frac{1}{2}$-foot zenith sector in the autumn of 1768, and corresponding ones taken with the two mural quadrants, revealed that zenith distances on Bird's mural quadrant were 1·1″ too small, whereas those on Graham's mural quadrant were 7·4″ too big. He consequently regarded these figures as the errors in the lines of collimation of these respective instruments, instead of dismounting them from the meridian wall and transposing them to face in contrary directions on the opposite sides of it. The great weight and size of those particular instruments would have made this a laborious task and incurred a considerable time-delay. An examination of the zenith distances of stars taken with Bird's quadrant in the autumns of 1765 and 1768 suggested that no appreciable change in the 1·1″ collimation error had occurred over the whole period from 7 May 1765 to 3 July 1772, when the instrument was disturbed to allow an achromatic object-glass to be substituted for the common one previously used.

Another error which was first detected by Maskelyne in zenith distance measurements with the new 10-foot zenith sector at St Helena was traced by him to an imperfection in the mode of suspension of the plumb-line (a fine silver wire) from the neck of the central pin around which the index arm of the instrument rotated. When he took the loop of the plumb-line off that pin, and then hung it on again, or when he replaced it by a new one, he discovered to his dismay that the other end of the plumb-line crossed the limb at a point 10″ or even 20″ from its initial position. He demonstrated this effect to a Royal Society committee at the British Museum on 11 September 1762, and ascribed it to friction which could be reduced but not entirely eliminated by making the neck of the pin thinner. He consoled himself with the fact that he had detected this fault, and made certain that it was not repeated by Bird when constructing a 6-foot zenith sector for Charles Mason and Jeremiah Dixon for their journey to Maryland and Pennsylvania for the purpose of settling the boundary line between these two provinces. Indeed, he referred to this new sector in 1768 as

> "the first which ever had the plumb-line passing over and bisecting a point at the centre of the instrument", and "so exact, that they found they could trace out a parallel of latitude by it, without erring above 15 or 20 yards".[4]

Maskelyne's interest in examining previous investigations of this nature by French and Italian astronomers, who had used instruments of a similar design, made him conscious of the uncertainties in Lacaille's measurements at the Cape, and those of Maupertius in Lapland made with another sector designed and supplied by Graham which was also employed subsequently for determining the length of a degree of the meridian arc between Paris and Amiens. It was when reading the joint account of Christopher Maire and Roger Boscovich (Jesuit Fathers) of their efforts to find the distance between Rome and Rimini, published at Rome in the form of an elaborate Latin treatise under the auspices of Pope Benedict

XIV in 1755, that he first became aware of an important non-instrumental reason for the plumb-line's departure from the vertical direction, besides that produced by the axial rotation of the Earth, with which Maupertius and others had been directly concerned. This was the lateral attraction upon the lead weight by mountains which ought to exist if Newton's postulate of universal gravitation were correct. Boscovich suspected that his result had been influenced by such an effect, but it was not an easy matter to confirm this conjecture.

At first it seemed that Mason and Dixon's measurements, made on level ground, provided a suitable standard of comparison, free from any suspicion of being affected in this manner. However, calculations made at Maskelyne's request by Henry Cavendish revealed that the correct value for the length of a degree could have been between 60 and 100 toises or roughly 500 feet greater than that actually measured, on account of the proximity of the Allegheny mountains to the west and the negative effect of the Atlantic Ocean to the east. Similar calculations confirmed that both Boscovich's and Lacaille's determinations could have been sensibly affected by the attraction of hills coupled with a defect in the attractions of the Mediterranean Sea and Indian Ocean respectively. Maskelyne made a comparison of Bird's 5-foot brass scale used by Mason and Dixon with Graham's brass standard at the Royal Society, and found the former to be slightly shorter, which affected the absolute accuracy of their result. Moreover, before it could be compared with the French estimates, it was necessary to make a further comparison between the Royal Society's standard and the two sent over from Paris by Jérôme de Lalande, which were both supposed to have been adjusted exactly to the length of the toise used by Charles la Condamine and Pierre Bouguer at Quito in Peru over thirty years previously. This was carried out by Maskelyne and Bird, and led to the result that: "the true length of the degree according to the Royal Society's brass standard, in the temperature of 62° of Fahrenheit's thermometer, is 363771 feet or 68.8960 English statute miles",[5] which was equivalent to 56 888 Paris toises on the Peru standard. Having thus obtained the relationship between these two units of measurement, it became possible to make an intercomparison of nine different determinations made during the years 1736 and 1768, from which Maskelyne concluded that either the meridians are not accurately elliptical or that the inequalities of the Earth's surface appreciably deflect the plumb-line, or both.

It was probably his reading of Bouguer's *La Figure de la Terre* (Paris, 1749) which made Maskelyne appreciate that the effect of the attraction of a mountain on the direction of a plumb-line had actually been experimentally investigated by the members of the Peru expedition, who found a mean deflection of 8″ using a $2\frac{1}{2}$-foot radius quadrant in the neighbourhood of Mount Chimboraco, which rises to about four miles above sea-level although only two miles above the general level of the province of Quito. The idea itself, of course, stemmed neither from Bouguer nor from Boscovich, but from Newton who had stated quite explicitly in the third book of his *Principia* (1687) "that a mountain of a hemispherical figure, three miles high and six broad, will not, by its attraction, draw the plumb-line two minutes out of the perpendicular".[6] Stimulated by the thought that confirmation of the French result would boost the faith of astronomers in the universality of gravitational attraction, and by Bouguer's express wish that such an investigation be made in England or France, Maskelyne began to wonder how and where this might be best achieved, and was instrumental in having a Royal Society committee established to consider this project. Charles Mason was given the not unpleasant task of touring the Scottish Highlands in the summer of 1773 with a view to recommending a suitable mountain on which the experiment might be conducted. He reported favourably on Schiehallion because of its height (=1081 m), tolerable detachment from neighbouring hills, steepness and comparatively narrow north–south base. This recommendation was formally approved and the experiment arranged for the following summer, supported by what was left of George III's over-generous fund for financing the 1769 Transit of Venus expedition.

Maskelyne himself was chosen as the most proper person to undertake this task, and had no difficulty in obtaining leave of absence from his Greenwich post while he did so. His late assistant, Reuben Burrow, went ahead to arrange that the temporary observing station and equipment, which had been shipped from London to Perth, would be transported safely overland to a prearranged location near the base of the mountain, where Maskelyne found them waiting for him when he arrived on 30 June. The major instruments used were the same 10-foot zenith sector with achromatic objective which he had used in St Helena (but now having a modified suspension for its plumb-line), and a 9-inch diameter theodolite made by Ramsden capable of observing angles to ±1′. He was also provided with two barometers made by Edward Nairne, a Gunter's chain, a long tape measure, two 20-foot long fir poles with stands to support them, and the 5-foot brass standard for adjusting them. It was while engaged in this work that Maskelyne had the insight that a better means of confirming whether his sector was truly in the meridian plane was to compare the difference in transit times for two stars of very different declination with their known difference in right ascension, rather than to find the clock-error from observations of the Sun's altitudes.

From a comparison of the results obtained from the astronomical observations and land-surveying operations carried out on Schiehallion during the three months from 20 July to 20 October 1774, full details regarding which were duly published in the *Philosophical Transactions* for the following year and need not be repeated here, Maskelyne formulated the following four conclusions:[7]

1. The attraction of Schiehallion was sensible in magnitude; hence every particle of matter is endowed with a gravitational force proportional to its mass.

2. This force varies, as Newton postulated, as the inverse square of the distance between any two attracting bodies, whether they be terrestrial or celestial.

3. The Earth's mean density is at least double that at its surface, implying a greater density nearer the core which result is "totally contrary to the hypothesis of some naturalists, who suppose the earth to be only a great hollow shell of matter".

4. In consequence of the deflection in the direction of a plumb-line produced by the less dense superficial parts of the Earth, errors were liable to be incurred in astronomical measurements of meridian arcs.

The first three of these experimentally derived results were undoubtedly of great interest and encouragement to the many contemporary proponents of the Newtonian philosophy, but it was the last which bore the most relevance to the subsequent refinement of astronomical science. The existence of errors in the length of a meridian arc value of the Earth's oblateness necessary for the computation of some of the lunar inequalities could still not be specified with certainty, which therefore constituted a limiting factor in the further improvement of the theory of the Moon's motion. Ironically, it was Pierre de Laplace's evaluation of terms in the variation of the lunar apogee and node which was ultimately to provide a welcome confirmation of the flattening of the Earth derived from the French expedition to Peru!

The trigonometrical surveys carried out in France had, of course, been undertaken not only for the purpose of determining the length of a degree of the meridian through Paris but also because of the general importance of such large-scale triangulation in preparing new and better maps for military and civil purposes. Thus in October 1783 the Comte d'Adhemar, French Ambassador in London, transmitted to Charles James Fox, one of H.M. Principal Secretaries of State, a memoir from Cassini de Thury setting forth the various advantages that would accrue if a series of triangles were carried from the neighbourhood of London to Dover and there connected to those already executed in France; claiming, in particular, that these combined operations would provide a more accurate determination of the relative situation of the world's two leading observatories at Paris and at Greenwich, than had hitherto been attained. It was this event which sparked off the Ordnance Survey of the British Isles.

At George III's command, Cassini's memoir was transmitted to Sir Joseph Banks, then President of the Royal Society, who in turn passed it on to Major-General William Roy, F.R.S., with the request that he organize the operation. On 16 April 1784, together with Banks, Henry Cavendish and Charles Blagden, Roy inspected Hounslow Heath and saw what had to be done to clear it before using it as a site for the measurement of a base line. Ramsden made a steel chain precisely 100 feet in length which was calibrated with the Royal Society's brass standard scale. After the results of the preliminary survey made with it had been published, Banks passed the memoir to Maskelyne for comment and opened a diplomatic correspondence with the Paris Academy of Sciences, while Ramsden began constructing the theodolite required for the angle measurements. The feature of Cassini's memoir which irritated Maskelyne was its author's assertion that these astronomically determined latitude and longitude differences were as much as 15″ and 11″ in error respectively. He believed that this would never have been stated, or indeed that the occasion for composing that memoir never have arisen, had Bradley lived long enough to publish his own observations; for then Cassini would have seen upon what foundation the latitude of the Greenwich Observatory had been established, and what differences of meridians between it and the other principal observatories of Europe had been deduced from various kinds of celestial phenomena.

Maskelyne believed that Cassini had been misled through his reading of a *Mémoire* by Lacaille (1755) where the conclusion drawn from a comparison of the differences in Bradley's zenith distances of 14 stars (published in 1752) and his own observations at Paris was that the latitude of Greenwich is 13″ to 14″ in excess of Bradley's value for it. This presupposed that Lacaille's own latitude for the Collège Mazarin (=48° 51′ 29·3″) and table of refraction were free from any error, and that both instruments measured angles correctly. A manuscript by Bradley given to him by Hornsby, coupled with his own twenty-two years of experience at Greenwich, gave him complete confidence in the reliability of the two mural quadrants there; but the sextant used by Lacaille had not been subjected to the same rigorous tests and there was good reason to suppose that most of the error lay in its construction. He knew, for instance, that Lacaille's observations yielded a latitude for the Paris Observatory 8″ in excess of the mean result of Maraldi, Le Monnier, and Cassini de Thury; and that the former's refractions were 2″ more than theirs at the altitude of the pole, or latitude at Paris (viz. 48° 57′ 4″). Besides, Lacaille's refractions were higher than G. D. Cassini's, Flamsteed's, Newton's, Bradley's, Mayer's, Simpson's; and even, as he perceived in another of Bradley's manuscripts that had come into his possession, than Lord Macclesfield's (whose 5-foot radius brass quadrant at Shirburn Castle had been rigorously re-examined).

These facts, and further comparisons between the other sets of data, left no doubt that the total arc of Lacaille's instrument was too large for the radius. It was pure luck that the errors compensated in such a way that his stellar declinations remained highly accurate and therefore agreed to within ±5″ with Bradley's, even though the refractions differed so much. Maskelyne perceived that Lacaille's refractions ought not to have been used to correct Bradley's observations, and insisted as a general rule:

> "that a table of refractions should be made *for every vertical instrument* from observations made with itself turned alternately north and south; and that the table, so made, applied to observations made with it, will give the true zenith distances, whether the total arc of the instrument be accurately just, or affected with a small error, or

148

however unequally it be divided below the pole, provided the divisions are equal between themselves in the part of the instrument lying between the equator, the zenith, and the pole".[8]

When the observations of each of the two astronomers in question were corrected by their own values for the refraction, the latitude of Greenwich was found to be only $4\frac{1}{2}''$ more than that established by Bradley and confirmed (to within $1''$ or $2''$) by Maskelyne in the course of establishing his catalogue of thirty-six principal stars.

As far as the longitude difference between Paris and Greenwich was concerned, Bradley's own value, based upon his observations of the eclipses of Jupiter's first satellite, was $9^m 20^s$, and he had inserted this value in the table giving the latitudes and longitudes of places prefixed to the posthumously published astronomical tables of Halley (1749). This figure had been accepted by English astronomers until James Short found $9^m 16^s$ from observations of four transits of Mercury in 1723, 1736, 1743, and 1753 made at Paris, London, and Greenwich; the latter determination being, in fact, the one adopted by Maskelyne in 1767 when preparing his edition of Mayer's tables (1770). His comparison of the data on eclipses of Jupiter's first satellite sent to him on request by Cassini de Thury's heir, collected by that astronomer at the Paris Observatory and independently by Messier at the Hotel de Cluny (2^s further east), with those amassed at Greenwich using the three telescopes mentioned earlier, yielded $9^m 30^s$ as the longitude difference of Greenwich from the Paris Observatory and $9^m 20^s$ from Messier's observatory. Since Du Séjour had also found $9^m 20^s$ from a solar eclipse in 1769, Maskelyne preferred this value "as being within a very few seconds of the truth"[9] until a better one should be derived from lunar occultations. His prediction that the intended trigonometrical operations might serve to determine the difference of meridians between Greenwich and Paris to great exactness without throwing any new light on the difference of latitude of those two observatories was fully confirmed. As his previous researches and expedition to Schiehallion had led him to anticipate, it was found that the result of Roy's triangulation gave a longitude difference of almost $9^m 20^s$—approximately 1^s less than it ought to have been on the assumption that the Earth's shape is spheroidal—but on no acceptable hypotheses regarding the figure and density distribution of the Earth could it be brought into satisfactory agreement with the accepted latitudes.

149

It seems astonishing that Maskelyne's astronomical prediction of the longitude difference between the two national observatories, announced several months before the trigonometrical operations were begun, should agree to the nearest second with the value published by Roy shortly after their completion. There is, however, a record by Joseph Lindley, Maskelyne's assistant from 1781 to 1786, of an independent chronometric determination of the longitude of Paris made from 20 September to 3 October 1785, yielding 9^h 19^m·8. This result was actually obtained by averaging the Paris–Greenwich time differences deduced from seven watches by John Arnold regulated to Greenwich Mean Time, according twice the weight to the three described as box watches (i.e. chronometers). The fascinating implication of this hitherto unrecorded fact is that Maskelyne sent Lindley across the channel on this 'chronometer run' in order to obtain a check upon the values found from astronomical observations and was in possession of this information *before* composing his critique of Cassini's memoir for the Royal Society! In these circumstances, therefore, there is less reason for astonishment and more for admiration at Maskelyne's initiative.

The involvement of the Royal Observatory with the testing of chronometers dates back to the beginning of Maskelyne's long period of office. After the commissioners for longitude had unanimously agreed that Harrison's fourth marine timekeeper (H4) had satisfied the most stringent condition imposed by the 1714 Act (cf. Chapter 4), they resolved that it should be subjected to a prolonged trial at Greenwich in the hands of the new Astronomer Royal in order to obtain more insight into the reliability of its mechanism under varying conditions of temperature and differences of position. This decision did not please Harrison, since Maskelyne was, in his opinion, a biased judge of his watch's merits. On most days between 5 May 1766, when it was set in motion and locked up in a box provided for it, and the end of its period of trial on 4 March 1767, H4 was wound up and compared with the transit clock at Greenwich by Maskelyne himself. At other times, this duty was performed by his assistant Joseph Dymond, and afterwards by the latter's replacement William Bayly, always in the presence of one of the officers of the Royal Hospital at Greenwich. To begin with, two or three comparisons were made during the day, but from 3 August onwards, with few exceptions, only one such observation was noted. During the first two months of the trial, the watch was placed horizontal with the face tilted upwards vertical, and horizontal with face downwards; then restored to its original position. Maskelyne's published report on this trial was most unfavourable. After giving an analysis of the gains and losses, and their daily variations, he concluded that H4 could not be depended upon to keep longitude to within 1° in a six-week West Indian voyage, nor to within ½° for more than a fortnight and even then only when kept at a temperature of a few degrees above the freezing point.

This verdict was completely opposed to its behaviour during its two sea-trials, but Harrison was not slow to point out good reasons for the apparently paradoxical result. Among these was Maskelyne's failure to make due allowance for its rate. Indeed, when this was done and the antics of H4 in different positions and extreme temperatures were neglected, its general performance was amply sufficient to justify the commissioners' views of its merits. The eventual recognition of Harrison's right to the second half of the maximum £20,000 reward and the payment of the rest of this sum to him in 1773, coupled with the success of Larcum Kendall's copy of it during Captain James Cook's second voyage to the South Seas, conclusively proved the practicability and general utility of the chronometer method for accurate longitude determination at sea. Thus all previous Acts relating to this problem were repealed, and a new Act (14 George III, cap. lxvi) formulated in 1774. The terms of this Act gave little encouragement to other watchmakers. The rewards were to be paid only for a timekeeper based on principles *other than* those which had now been made public and their amounts were halved; moreover, they were not to be paid until *two* such timekeepers had first been tested for twelve months at the Royal Observatory, then in *two* voyages round Great Britain in opposite directions (and possibly other voyages), followed by a further trial at Greenwich. The principles and practice of construction, together with descriptions, theories, and explanations, had all to be made public and both instruments handed over to the Board of Longitude. A *two-thirds* majority, not merely a simple majority, of the commissioners had finally to indicate their approval.

Among the bold pioneers of chronometry with whom Maskelyne became involved in prolonged disputes was Thomas Mudge, who had watches officially tested at the Royal Observatory on three occasions between 1779 and 1790 and just failed to qualify for the parliamentary award. After the third attempt had been judged unsuccessful, Mudge's son published a pamphlet in which, among other things, he openly accused Maskelyne of deliberately falsifying the registers containing the accounts of these trials. This libellous attack provoked a reply from the latter, which caused his accuser to write a second pamphlet—all three being published during the year 1792. The most significant of Mudge's numerous objections was his criticism of the criterion which Maskelyne had adopted for assessing the accuracy of the watches' performances; namely, that of basing the mean daily rate upon the first month's observations and applying the value thus obtained to the whole of the following year. There was no particular reason for this procedure, which turned out to be more stringent than those which had been applied to Harrison's watch. The fundamental problem lay, of course, in the ambiguity of the words 'accuracy' and 'error' when applied to chronometers. The matter was taken up by a House of Commons Committee which decided to award £2500 to Mudge's father although its sub-committee's report stressed that

> "no judgment can be formed of the exactness of any time-keeper by theoretical reasoning upon the principles of its construction, with such certainty as with safety to be relied upon, except it be confirmed by experiments of the actual performance of the machine".[10]

Another watchmaker who had been making chronometers deemed by a committee of the Board of Longitude to be 'incomparably better'[11] than those of Mudge, was Thomas Earnshaw. Earnshaw introduced two very important technical improvements in design—the detached escapement and compensation balance—and the results of three sea-trials of his chronometers, in competition with others by Mudge, John Arnold, and other makers, during the years 1796–98, were decidedly in his favour. He also submitted two watches for trial at the Royal Observatory on three occasions and *just* failed on the basis of Maskelyne's criterion to satisfy the requirements of the 1774 Act. Only after a bitter dispute with Sir Joseph Banks was he finally awarded £2500 for his praiseworthy efforts, thanks partially to Maskelyne's intervention in his favour.

A consequence of the experience gained from these events, and from the various trials held at the Royal Observatory, was the establishment by the Admiralty of a consistent method of chronometer rating. At first, during the period of the so-called 'premium trials' (1823–35), prizes were awarded to those who deposited chronometers which kept time to within a previously specified Trial Number found by taking twice the difference of the greater and lesser mean monthly rates and adding this quantity to the mean of the monthly extreme variations. Thus the smaller the number, the better the chronometer. Of all those submitted, that which had the lowest trial number (=2·27 in 1829) was made by E. J. Dent. A modification of this plan was introduced in 1840 by the seventh Astronomer Royal, George Airy, who preferred to define the trial number as $a+2b$ where a is the difference between the greatest and least weekly sums, and b is the greatest difference between the sums of two consecutive weeks. This system of rating, which continued to be used until the outbreak of World War I (1914), made it possible for a machine which alternately gains and loses to have a very small rating number.

In our brief survey of Maskelyne's conscious endeavours to revive British astronomy, we have seen how well he succeeded in placing the affairs of the Royal Observatory upon a firm administrative footing and how, aided by an effective well-disposed Board of Visitors, he ensured that the instruments and buildings under his care were maintained in a proper state of repair. He wisely adopted the observational policy of ignoring cosmological speculation and concentrating his attention on determining, to the highest possible degree of precision, the equatorial coordinates of only thirty-six principal stars, rather than attempting to follow in the footsteps of his predecessors and produce an extended star catalogue. It was at Maskelyne's instigation that a Royal Warrant was signed in 1767 authorizing the printing of future Greenwich observations and their payment by the Board of Ordnance provided that the cost did not exceed £60 p.a. The regular publication of these began in 1776— over one hundred years after the Observatory's foundation—when the data for the years 1765 to 1769 which he had conscientiously been submitting to the Board of Visitors at regular intervals in accordance with the 8th draft (actually the 9th) regulation were printed in a single volume. Three later volumes containing the observations from 1775 to 1786, 1787 to 1798, 1799 to 1810 were published in the years 1787, 1799, and 1811 respectively. Initially, only 25 of the 500 copies printed were presented within Great Britain and Ireland, 11 to European Universities and academies, and 2 to the American colonies (for the libraries of Harvard College and the Philosophical Society of Philadelphia). Other institutions were subsequently added to the list.

While thus promoting the cause of astronomical science through his routine work at Greenwich, Maskelyne never lost sight of the principal object of his observatory's existence; namely, the improvement of navigation. The stars which he selected for his observing programme were of great use to the practising navigator when taking sightings of lunar distances. His various improvements to nautical instruments such as the Hadley quadrant and development of the allied observational techniques; his editing of Mayer's lunar theory and tables; and his testing of chronometers, were all directly relevant to the needs of the British mariner. Lastly, and principally, it was as a result of his initiative and skilful organization that the *Nautical Almanac* and "Requisite Tables" for compensating the effects of refraction and parallax became for a century or more indispensable aids to finding longitude at sea by the method of lunar distances. Because of the utility of publishing the almanac two or three years in advance, Maskelyne was empowered by the Board to engage as many additional staff as he found necessary for this purpose, who were able to work in their own houses. It seems that, throughout his lifetime, the total seldom, if ever, exceeded four computers and a comparer. His insistence on the independence of the results is illustrated by his instant dismissal of two nautical almanac workers called Keach and Robbins, when he found them copying each other's calculations.

An unexpected increase in the computational load for this annual publication followed William Herschel's discovery of Uranus in 1781. The celestial longitude of this new planet was calculated for the 1st and 15th days of each month, and the results subjoined to the Nautical Almanacs from 1787 to 1795. In later years these computations were extended, along with those of Jupiter and Saturn, to three days (viz. the 1st, 11th and 21st) of each month. It would appear that this was when a fourth computer had to be engaged to help lighten the individual labour, and as an additional encouragement to those involved in this laborious occupation their salaries were increased from £80 to £100 p.a. Two smaller increases of £5 p.a. had, in fact, been previously awarded in 1767 and 1781. This mark of appreciation must have spurred the three gentlemen and one lady concerned (Messrs Henry Andrews, Thomas Brown, Nicholas James, and Mrs Mary Edwards) to greater efforts, since within the next four years they managed to calculate the tables and ephemerides ten years in advance, and thereby worked themselves out of their jobs. The Board of Longitude awarded them compensation, and they were given the alternative employment of calculating the eclipses of Jupiter's satellites from 1795 to 1804 using new tables published by Lalande in 1792. The importance of this development became evident when the comparer, the Reverend Malachy Hitchins, discovered nine folio pages of errata in the previously computed tables of these satellites.

This experience led to the adoption of Lalande's tables as a basis for the computations of future almanacs. The resulting increase in the accuracy of the solar ephemerides (to the nearest 0·1″) and of the tables of primary planets and eclipses of Jupiter's satellites (to within 1″) involved even more work, although a slight reduction in labour was introduced in 1809 by the decision to restrict computations of lunar occultations to stars of the first and second magnitudes only. That same year, new and more accurate solar and lunar tables compiled by Joseph Delambre and Johann Tobias Burg were presented to Maskelyne by the French Board of Longitude but, as a result of a temporary setback following the death of the experienced comparer, Malachy Hitchins, the resulting improvements were first incorporated in the *Nautical Almanac* for 1813. Thanks to the successful representations made by Maskelyne to the Board of Longitude on behalf of the *Nautical Almanac* workers, a computer was now receiving more than double the salary that had been approved twenty years before—namely, £205 instead of £100 p.a., on completion of one year's almanac—while the comparer's salary had risen to £250.

These sums were far in excess of the annual salary paid to the Astronomer Royal's assistant which for the first hundred years of the Observatory's history officially had been a mere £26, though Maskelyne had supplemented this small sum by a further £34 out of his own pocket. But, as he pointed out in a memorial to the Lords of the Treasury (1771), even £60 p.a. was still inadequate. Three years later the assistant's official salary was raised to £86 p.a. and, less than a year before his death, Maskelyne made a further representation to get him more than £100 p.a. But, his salary aside, a melancholy picture of the life of this unfortunate individual at the Royal Observatory is painted by John Evans in the *Juvenile Tourist* of 1810, where he remarks:

"Nothing can exceed the tediousness and *ennui* of the life the assistant leads in this place, excluded from all society, except, perhaps, that of a poor mouse which may occasionally sally forth from a hole in the wall, to seek after crumbs of bread dropt by his lonely companion at his last meal! This, of course, must tend very much to impede his inquiring [*sic*] astronomical information, and damp his ardour for those researches which conversation with scientific men never fails to inspire. Here forlorn, he spends days, weeks, and months, in the same long wearisome computations, without a friend to shorten the tedious hours, or a soul with whom he can converse."

Small wonder, then, that Maskelyne should have had no fewer than 25 different assistants; and that his successor John Pond, while doing all that lay in his power to make his own assistants' lives bearable, should have privately regarded them as mere drudges.

NOTES

1. Quoted from the Xerox copy of R.G.O. MS. 617 preserved at the National Maritime Museum Greenwich, pp. 56–57.
2. *Ibid.*, pp. 72–74.
3. *Op. cit.*, p. v.
4. Maskelyne, 1768 c, p. 271.
5. Maskelyne, 1768 d, p. 325.
6. Quoted from Maskelyne, 1775 a, p. 496.
7. Maskelyne, 1775 b, pp. 532–534.
8. Maskelyne, 1787 a, p. 180.
9. *Ibid.*, p. 186.
10. Mudge, 1799, p. xxxii.
11. R.G.O. MSS. 534, Board of Longitude Confirmed Minutes, 6 (11 June 1796), p. 265.

8

An Age of Expansion

WHEN John Pond took up his appointment as sixth Astronomer Royal on 13 April 1811, he found himself in a very fortunate position. He stepped into the Royal Observatory as a 'going concern', with the manuscript observations of Flamsteed and Halley deposited under his care and the entire legacy of Bradley's, Bliss's and Maskelyne's respective contributions to the Greenwich data now finally in print. An unimportant exception was the missing fifth observation book of Halley, already referred to in Chapter 4. Pond was fortunate, too, in another respect: he fell heir to a new instrument, a 6-foot diameter mural circle, which had been ordered for Greenwich on the recommendation of a Royal Society committee in 1807 after hearing from Maskelyne that most foreign observatories were already equipped with a divided circle which was intrinsically more reliable than a mural quadrant for zenith distance measurements. It would appear that what the latter originally had in mind was a 4-foot diameter vertical circle *movable in azimuth*, like the magnificent 5-foot diameter vertical circle constructed by Ramsden in 1789 for Guiseppe Piazzi in Palermo; but the committee decided in favour of a fixed instrument of still greater radius. Before his acquisition of Ramsden's instrument enabled him to rely entirely upon his own observations, this eminent Italian astronomer had, in fact, been using Maskelyne's thirty-six stars as fundamental reference points in the construction of his first catalogue of 6748 stars published in 1803. Another motivation for Maskelyne's request for a circular instrument was that he had reason to believe that the wear of the central cylinder of Bird's quadrant had begun to produce significant errors in zenith distance.

During the course of the next five years the London instrument-maker Edward Troughton was busy making the Greenwich mural circle, which he brought to the Royal Observatory on 23 May

1812, where it was soon to become a vitally important tool in Pond's observing programme. In the interim, however, the latter had continued to employ Bird's mural quadrant to observe those stars which he intended to observe afterwards with the mural circle. This enabled the nature of the derangement already experienced by Maskelyne with the former instrument to be more accurately ascertained. His investigation led Pond to conclude that the errors had occurred gradually, having been greatest around 1800. Single observations could differ from the mean of a great number by as much as 5″, the amounts depending upon the part of the limb at which the measurements were made. He believed that part of the uncertainty in previous values had arisen from the conventional practice of adjusting the instrument by passing a plumb-line over two points, since one point may be deranged by a partial expansion and the whole instrument moved in consequence to a new position that could well be less exact than the initial one. Here, then, is another disadvantage of a plumb-line instrument, to add to that already mentioned in Chapter 7, for which it is equally difficult to make a quantitative correction. The only way to eliminate such uncertainties was to estimate the position of an instrument from the astronomical observations themselves.

The great advantage of using the mural arc was that its principle—essentially that of a vertically mounted theodolite—enabled the observer to determine the total arc between a star (or the Sun) and the pole without any intermediate point, and thus dispense with the need for a plumb-line. The co-latitude, or distance between the zenith and the pole, therefore became more a question of curiosity than of absolute necessity; or, expressed differently, the results were entirely independent of the place of observation. From any mean point, very close to but not exactly coincident with the zenith, Pond had simply to measure the star's (or Sun's) distance southwards and the pole's distance northwards, to find the apparent north polar distance. The position of the pole on the instrument was ascertained by numerous observations of circumpolar stars (e.g. 120 on Polaris, 68 on γ Draconis, etc.) made during the last six months in 1812, the corrections for refraction being based on Bradley's formula. The index error of the instrument for the day of observation was ascertained by comparing the observed places of 30 bright stars with those so accurately determined by Maskelyne. This quantity was then applied to all the observations to obtain the true polar distances. Pond later summarized the results of these first few years' work as follows:

> "For the first two years after it was erected this Instrument was most assiduously employed in the formation of a Standard Catalogue of stars few in number, but whose places were determined with a degree of precision I believe never before attempted. It was intended to use these as points of reference in the determination of other stars, precisely in the same manner as has been done by Dr Maskelyne with his catalogue of right ascensions. When it appeared to me almost hopeless to improve this Catalogue by continued observation I devoted the Instrument for the next three years to the formation of a more extended catalogue of 300 stars deduced entirely from comparison with the first. A less scrupulous method for observing was in this case therefore adopted, but quite exact enough for the object I had to accomplish. I have now every reason to believe that with the standard catalogue above alluded to & the deduced catalogue were exact within the narrow limits I supposed indeed many circumstances now lead me to conclude that they are even more exact than I myself conceived them to be at the time of observation but during the whole of this period the Transit Observations were entirely neglected I mean that no attempt either was made or *could have been made* to have obtained any additional knowledge with respect to the Right Ascensions of the fixed stars."

What had prevented Pond from simultaneously carrying out transit observations was the need for him to work on the new instrument together with his assistant Thomas Taylor or his ward John Belleville (who later changed his surname to Henry). A second assistant was certainly required if the right ascensions of the stars were to be simultaneously determined, for six microscopes (attached to the wall rather than the circle itself) had all to be read and the results for the north polar distance of each star averaged before the corrections for refraction could be applied. He twice appealed to the Board of Visitors before such help was forthcoming, paid for by the Admiralty instead of the Board of Ordnance out of money set aside for naval services. The help of a labourer at 18 shillings per week was also financed from this same source. The decision to raise the first assistant's salary and introduce the post of second assistant coincided with the Observatory's acquisition of a new 10-foot long transit instrument costing £315, which had also been constructed by Troughton. Henceforth, throughout a further period of almost five years, no observations of any importance were made with the mural circle, since the whole attention of the Astronomer Royal and his assistants was directed to the investigation of the right ascensions of the thirty-six stars in Maskelyne's catalogue.

These quantities had originally been found from numerous observations with the previous transit instrument of each star's *difference* in right ascension from α Aquilae, added to the absolute value of that particular star's right ascension. Maskelyne's method for discovering the latter had been to adopt, but only provisionally, Bradley's determination of it. Then, from observations of the transit of the Sun and these other stars in spring and autumn, he inferred the right ascension of each with respect to the Sun. From the Sun's observed meridian zenith distances on Bird's mural quadrant, corrected for refraction, parallax, and collimation error, using Bradley's values for the obliquity of the ecliptic and the latitude of the Royal Observatory, he computed the Sun's right ascension and declination. The Sun's right ascensions found from the transit observations would be expected to contain an error due to an error in the assumed right ascension of α Aquilae that would remain constant throughout all seasons of the year. However, the Sun's right ascensions calculated from the observed zenith distances would be affected in contrary directions at opposite seasons of the year on account of the unknown errors in the applied refractions, parallaxes, obliquity, and latitude; thus the mean of two corrections of the Sun's right ascension at the vernal and autumnal equinoxes would be the true correction of the assumed right ascension of α Aquilae. The difference of the two independent corrections would eliminate any error in obliquity and yield the errors in the computed declinations—provided that the refractions were accurate. The disadvantages of this ingenious method were that *two* stars, α Aquilae and one other, had to be observed on each occasion, thereby incurring the possibility of a double error; and, more significantly, that a failure to observe α Aquilae (owing to bad weather or the inconvenient hour at which it crossed the meridian) would render the observation of all the other stars entirely useless. Hence many years had been needed to produce a catalogue with the accuracy required.

Pond's reconstruction of Maskelyne's catalogue was based upon principles which, in his own opinion, were entirely new and of unrivalled precision; moreover, they enabled the same results to be obtained more quickly (e.g. in one year rather than in three). The key innovation was that *every* star, and not just one, was chosen as a point of reference for all the others. Their relative distances on the equator or meridian had first to be found, then a common point of departure selected for the purpose of standardization. Just as the index error of the mural circle could be used as the point of departure in the measurement of the north polar distance, so could the clock-error found from each star's observation be applied to a star for which the right ascension is supposed known. The absolute right ascension was then obtained in the manner described above and the result recorded. Pond found, in fact, that the right ascension which he derived for α Aquilae by this sophisticated means agreed to within $0^s \cdot 01$ with that found by Maskelyne; and, in general, despite the unparalleled excellence of his new mural circle and transit instrument, he was unable to find errors greater than $0^s \cdot 1$ in Maskelyne's catalogue. Local refraction imposed an insurmountable limit to the declination observations, but there was no such restriction on the values of the right ascension.

A puzzling phenomenon affecting the declinations of several principal stars, detected by the Reverend John Brinkley of the Dunsink Observatory, Dublin, using an 8-foot diameter altazimuth instrument designed and partially constructed by Ramsden on the same principles as that previously supplied to Piazzi, was tentatively attributed by Pond to the effect of seasonal temperature changes upon the refraction. He rejected Brinkley's own interpretation of the observed small periodic changes as being due to parallax, for the simple reason that he was unable to deduce analogous results from his own mural circle observations. Although astronomers, largely as a result of Bradley's discovery of stellar aberration (cf. Chapter 5), had all come to accept the idea of the Earth's revolution about the Sun, the problem of quantifying the annual stellar parallax and thereby setting a scale on the Universe was one of great practical importance which Pond felt ought to be closely studied. Since he did not want to interrupt his observing programme with the mural circle in order to prosecute such an investigation, he proposed to the Board of Visitors in 1816 that two or more telescopes fitted with micrometer eyepieces should be fixed to stone piers and directed towards the stars whose parallaxes Brinkley had suspected. His suggestion was accepted, and two 10-foot telescopes were attached securely on the circle and quadrant piers and then pointed towards α Aquilae and α Cygni respectively. He observed each of these and another star differing little from it in declination but by several hours in right ascension since this procedure happened to be suitable for detecting a parallax effect. However, the results of those observations merely confirmed his view that parallax was *not* responsible for the discrepancies to which his Irish colleague had drawn attention. We can now, with hindsight, ascribe it to the so-called 'variation in latitude' caused partially by the annual displacement of water masses between the arctic and antarctic regions, by melting and freezing, producing an annual oscillation of the terrestrial poles; but this came to be recognized only towards the end of the nineteenth century from entirely independent considerations.

The same method could not, however, be used in the case of a star such as α Lyrae for which there is no other star of nearly the same altitude (or declination) but opposite in right ascension and sufficiently bright to be observed throughout an entire year. Since Brinkley's Dublin observations had shown the parallax of γ Draconis to be negligible, Pond used the mural circle to measure the difference in parallax between the star and α Lyrae, hence the absolute parallax of α Lyrae. The technique which he applied in this investigation was, however, completely novel—involving a comparison of measurements made by direct vision with others made at other transits after the star's image had been reflected off a mercury surface. Previously, he had found difficulty in masking this artificial horizon from the effect of draughts, but experiments carried out during the last three months of 1819 had proved successful. It follows from the principle of equality between the angles of incidence and reflection that half of the angle contained between the two readings on the circle is equal to the altitude of the star, while the reading at the point of bisection indicates the horizontal point. Knowing the latitude of the place of observation, the north polar distance was then immediately determined. The results of these new measurements were negative, thus Pond concluded that what others such as Piazzi and Calandrelli in Italy believed to be parallax were really nothing more than imperfections in the construction of their instruments.

A disadvantage of the method just described was that since the circle was fixed in the meridian, twenty-four hours or frequently (on account of cloudy weather) several days would elapse before a complete observation of the double altitude could be obtained. The only way to remedy this was to employ two mural circles simultaneously to observe a star on the same night, in the one case by direct vision and in the other by reflection. Thus a new and preferably identical instrument was required. The four reasons given by Pond at a Royal Society Council meeting on 23 June 1823 for having a second mural circle installed at the Royal Observatory, are

"1. If one instrument should be under repairs the observations might be continued with the other without interrupting the regular duties of the place.

2. The Instruments will continually act as a check upon each other, and indicate any derangement that may take place in either.

3. When the two instruments are used together and in the same manner we may be able more exactly to estimate how much of the occasional discordance attending the most careful observations is to be attributed to accidental irregularity of refraction, and how much to the Instrument.

4. But the most important improvement will no doubt consist in being able to observe the same star at the same time both by direct vision and by reflection, and by reversing the operation alternately with each instrument, it seems difficult to limit the accuracy that may be ultimately obtained; four systems of divisions are thus brought to bear upon the measured angle, one half of which only is required to form a result."[1]

Points 3 and 4 are covered by what has been said already. Pond's concern with the possibility of derangement and repair had its grounds in his discovery of errors in his circle which appeared to vary from about 2″ at the Equator to between 6″ and 8″ near the horizon. He had first detected these in the autumn of 1819, and traced their cause to the rupture of screws fastening the braces holding the telescope tube to the circle. After Troughton had applied additional braces the observational accuracy had been restored.

The onus of paying the estimated cost of £800 for this new instrument had fallen upon the Lords of the Admiralty. They, at Pond's suggestion (made primarily on the grounds of administrative convenience), had now accepted full responsibility for the affairs of the Royal Observatory. This marked the end of the long-standing association with the Board of Ordnance; but the change was highly desirable since that Board had become increasingly less competent to assess the importance of the work being done at Greenwich and the nature of the computational problems then facing those responsible for the compilation of the *Nautical Almanac*. Under the terms of the Act 58 George III, cap. xx of 1818 "for more effectually discovering the Longitude at Sea ...", the constitution of the Board of Longitude was radically revised and the versatile scholar-scientist Dr Thomas Young appointed as its paid secretary at the salary of £300 p.a. He was also given the additional responsibilities of superintending the regular publication of the *Nautical Almanac*, and looking after the regulation of chronometers, for which he received two further increments of £100 p.a. Although the recording of the rates and the testing of these refined pieces of mechanism which were being offered for government purchase had been imposing an unwanted additional burden on Pond's staff, this service was unquestionably of great value to the navigational aspect of the Observatory's work and was to be encouraged now that it had become such a dominant feature of that institution's activity. The Greenwich time-ball, erected in September 1833, is symbolic of this same interest, since its function was to enable ships' captains on the Thames to check the errors and adjust the rates of their own chronometers before setting out to sea.

A memorable and far-reaching proposal made by Davies Gilbert (formerly Giddy), seconded by Sir Joseph Banks, at a meeting of the re-constituted Board of Longitude on 3 February 1820, was that an observatory should be founded at the Cape of Good Hope. This idea was subsequently taken up by a committee for examining instruments and proposals which recommended that an astronomer be sent there with portable instruments for site testing. The Foreign Secretary approved the suggestion, instructing the Governor at the Cape to allot a suitable piece of ground for that purpose at the expense of H.M. Colonial Government and to lend every assistance to the astronomer appointed. The King sanctioned the Board's decision to select three persons to pioneer this project, namely the astronomer, his assistant, and a 'labourer' (in effect, a second assistant), and to pay them annual salaries of £600, £250, and £100 respectively.

The site chosen by H.M. First Astronomer at the Cape (the Rev. Fearon Fallows) for his observatory was on Tiger Hill, three miles from Cape Town. Before the end of 1823—little more than two years after he began observing there—Fallows sent back a preliminary catalogue of 273 southern stars, made with a small transit instrument by Dollond and an indifferent altazimuth by Ramsden. The results were later presented to the Royal Society under the title "A Catalogue of nearly all the Principal Fixed Stars between the zenith of Cape Town, Cape of Good Hope, and the South Pole, reduced to the 1st of Jan. 1824".[2] The permanent instruments and the official sanction of his designs for the observatory reached him at the end of that year. They might have done so earlier if Pond had not decided to keep the mural circle, originally ordered in 1820 from Thomas Jones of Charing Cross for the Cape Observatory, for his own observatory at Greenwich. Fallows received a second copy. In the first place, since it was made as a replica of Troughton's instrument, it was ideally suited to Pond's purpose (cf. above). Secondly, the Astronomer Royal was anxious to acquire it as quickly as possible to make simultaneous observations of the right ascensions of those principal fixed stars which an investigation in 1882 had shown to exhibit anomalous changes in declination that could not be attributed to systematic instrumental errors, since those should have been detected and eliminated by the newly adopted procedure of comparing direct and reflected images. The effect was such that the majority of the stars tended to be south of the places predicted from a comparison of Pond's own 1813 catalogue with Bradley's 1756 one, after due allowance had been made for precession. The greatest deviations were found in the very bright stars Capella, Procyon, and Sirius; Castor, α Aquilae, and α Cygni were, on the other hand, notable exceptions to the general rule.

An essential prerequisite to the acquisition of this other mural circle, which was finally installed at Greenwich early in 1825, was that two more assistants be hired in order to operate it. Accordingly, Pond began in 1823 to make representations to the Lords of the Admiralty for such additional help, plus better financial conditions for those currently employed. His first assistant, Thomas Taylor, was actually enjoying the maximum salary of £300 p.a. permissible under regulations formulated five years earlier, although the normal arrangement was intended to be £200 p.a. at the beginning of service, augmented by £20 p.a. after every three years until fifteen years had been completed. His second assistant, John Belleville (*alias* Henry), had already completed eleven years on a scale beginning at £100 p.a. with a £10 p.a. increment after every three years; thus he was then receiving £130 p.a. The matter was viewed favourably by the Admiralty, who obtained H.M. Treasury's approval to revise these scales (making them applicable to four assistants) as follows: 1st assistant begins at £300 p.a. + £10 p.a. up to a maximum of £500 p.a. The 2nd, 3rd, 4th assistants would each receive £100 p.a. +£10 p.a. up to a maximum of £300 p.a. It was agreed that the Astronomer Royal's appointment need not be made from within these ranks, and that advancement to first assistant should be by merit and not by seniority.

Those conditions were not, however, intended for the existing assistants, since, in Pond's opinion, Thomas Taylor was incapable of conducting the business of the Royal Observatory upon an enlarged scale. An agreement was reached in 1824 that the latter should be allowed to retire in due course with a suitable pension, and be replaced by a new and superior assistant. At about the same time his son Henry Glanville Taylor and William Richardson, who had already been employed by Pond in a supernumary capacity, were officially appointed to the newly created posts of third and fourth assistants. Further help was volunteered by Captain William Ronald, who had heard of the need for an assistant at the Cape Observatory and indicated his willingness to be considered for such a post either there or at Greenwich. Pond was able to write a strong recommendation of this gentleman's fitness to assist the astronomer at the Cape. It was, in fact, Ronald who brought the second copy of Jones's mural circle out to Fallows late in 1826, and who helped him to erect it there.

At about the same time (1825) as he acquired Jones's first circle to enable him to pursue his policy of making measurements of stars' right ascensions, Pond became the target for a sharp criticism by Stephen Lee, then Assistant Secretary to the Royal Society. Lee's situation placed him in an excellent position to consult the original manuscripts of the Greenwich observations in the Royal Society's archives, and to compare these with the published versions— permissible under the terms of George IV's Warrant of 19 May 1820 which stated that the former were to be "as accessible to public inspection as is consistent with their safety & the uninterrupted performance of the duties of the Observatory".[3] Lee submitted two papers for publication in the *Philosophical Transactions* regarding some 350 contradictory statements which he had found in the Greenwich observations for 1821, contained in the fourth printed volume issued under Pond's name in 1824. The President of the Royal Society, Humphry Davy, who had been a lifelong friend of the Astronomer Royal, was not keen for such matters to be aired in public, and persuaded Lee to address them to the Royal Society

Council instead; thus they are to be found only by consulting that Council's manuscript Minutes.[4]

Lee distinguished eight different classes of errors in those observations:

1. Inaccuracies in the means of two or more microscope readings (60 in all).

2. Contradictory readings of two microscopes (64 in all).

3. The barometer and thermometer readings in the two journals containing circle observations (286 disagreements).

4. Differences in transit times of two stars of different declinations, observed on the same night with the transit instrument and the circle, revealed that the plane of the circle did not quite coincide with the meridian.

5. The north polar distances of γ Draconis and other stars on successive nights, and their differences, should reveal any inconsistencies due to refraction, proper motion, precession, aberration, and nutation. In fact, they revealed "the most strange & unaccountable anomalies" which implied that the observations may have been inaccurately made or carelessly registered, or else that the instrument lacked stability.

6. The intervals between the principal wires of the transit instrument—there were now seven of these—should have remained equal for any given star, but were not.

7. In one case, decimal divisions had been mistaken for sexagesimal parts of a second.

8. The differences between the readings of one microscope (A) with those of the five others (B,C,D,E,F) for the same star observed on several successive nights and for different stars observed on the same night, ought to have been constant for all positions of the telescope unless there were inequalities in the divisions of the circle or imperfections in the centring of it. In order to eliminate these two possibilities, Lee had compared only those observations where the telescope was directed very near to the same point in the heavens. The fact that he still found considerable anomalies meant that the observations themselves must have been inaccurate.

Supplementary examples extracted from the 1817 observations in the third of Pond's printed volumes appeared to confirm that this last fault lay with the observer and not with the instrument; whereas a cursory examination of the 1822 observations in the fourth volume provided further support for the first six points. Lee also considered that he had found evidence for badly adjusted microscopes and defective micrometer screws. The conclusions which he drew from his detailed analysis of these data are stated as follows:

"1. It appears that out of 3095 observations with the Mural Circle entered in the separate journals for two & six microscopes 1258 contradict each other in respect to the state of the Barometer on[e] or both Thermometers or all three of them.

2. That out of the selfsame observations 81 are found to disagree in the readings with Microscope A or B and sometimes with both.

3. That out of 758 observations made by Reflection and read off with six Microscopes there are 109 errors in the computation of the mean readings.

4. That in 4124 Observations in the Journal for the Microscopes there are 109 errors in the computation of the mean readings.

5. That the plane of the Mural Circle seems never to have been very accurately adjusted to the plane of the meridian; though this deviation is not so great as to cause any very considerable error in the North Polar Distances.

6. That the differences between the readings with Microscope A and the other Microscopes seem to show that the Observations with the Mural Circle are subject to an uncertainty of several seconds.

7. That the Observations with the Greenwich Transit do not appear to exceed in accuracy what may be obtained from an Instrument of much inferior size."[5]

Lee concludes by remarking that the manuscripts submitted to the Royal Society were, in his opinion, not the original observations; and hoped that if this were so, the latter ought also to be deposited so that comparisons could be made.

Lee was undoubtedly justified in drawing these conclusions from his study of the internal evidence in the Greenwich observations. Pond's main line of defence was that he had been over-worked and under-staffed. After September 1821, he had given priority to the need for a complete revision of his standard catalogue of north polar distances and treated the transit observations as being of minor importance, especially since the pivots of the instrument had begun to wear and required to be repaired by the maker. The method of observing both directly and by reflection with Trough-ton's mural circle had added to his work-load, and to make matters even worse his first assistant had been laid off for several months with a severe illness. He ascribed many of the contradictory readings of the same microscope to printing errors and his own "little want of vigilance as an editor", whereas regarding the thermometer and barometer readings "almost every case of alleged contradiction will be found to arise from not having copied into one Journal some small change indicated in the other".[6] Many of the errors which Lee had marked in red ink on a copy of the fourth printed volume of Greenwich observations were only $\pm 0''\cdot 1$—inevitable in taking mean values. Only by immensely increasing the number of observations could uncertainties of this order be made to disappear. The verdict of the Royal Society Committee appointed to examine this matter was to exonerate Pond from any "culpable inattention to the immediate duties of his important office"[7] and to blame his first and second assistants for neglecting to record the barometer and thermometer readings, while rejecting Lee's claim that the mural circle had deviated from the meridian.

Had the matter been allowed to rest at this point there would be no need to suppose that Lee had an 'axe to grind', or that the motivation behind this attack on the Royal Observatory's observations was anything more than his dedication to the cause of scientific truth. Immediately, however, he brought forward a fresh set of charges against the low quantity, as opposed to quality, of the eclipses of Jupiter's satellites and cometary sightings recorded by the Greenwich observers during the years 1821 and 1822. The Royal Society's committee regarded bad weather conditions as a sufficient explanation of why only 16 out of 78, and 33 out of 89 such eclipses predicted in the *Nautical Almanac* for 1821 and 1822 respectively had actually been observed by Pond and his assistants. On the other hand, it supported Lee's objection that no observations had been made on a comet seen elsewhere in Europe from 21 January to 7 March 1821, nor on two more comets seen elsewhere from 15 May to 30 June and 13 July to 22 October 1822. These had, of course, no bearing on the subject of astronomical navigation which was the prime concern of the Observatory. Lee also drew attention to the facts that on only one night in six throughout 1821 had observations been made at Greenwich, and that there had been only one assistant at the transit instrument. These additional objections were, however, entirely explicable in terms of the inclement weather, Pond's policy of concentrating his attention upon the mural circle observations, and the long illness of his first assistant (leaving only the second assistant to operate the transit). Predictably, therefore, the Admiralty's reaction was not to censure the Astronomer Royal, but to acquire him even more assistance than the staff of four which he now had under his charge—a measure rendered all the more necessary for the efficient working of the Royal Observatory now that Jones's circle had come into regular use early in 1825.

There was, however, a decided ambiguity between the type of assistant whom the Admiralty wished Pond to employ at Greenwich, and the persons whom he had actually engaged. The Admiralty's policy of increasing the financial status and promotion prospects of the Astronomer Royal's staff was adopted with a view to encouraging a young gentleman with a university education to apply for the first assistant's post whereas, as Pond confided to Dr W. H. Wollaston:

> "I want indefatigable hard working & above all obedient drudges . . . men who will be contented to pass their day in using their hands & eyes in the mechanical act of observing & the remainder of it in the dull process of calculation. Persons of such habits & who will limit their views to such occupations I do not expect to find in the highly educated class of our Universities or should they be found I should feel some compunction in consigning them with their acquirements to such forbidding task work. For as to Mathematical talent of a superior order & the power of inventing formulae it is not what we have occasion for and would be thrown away."[8]

This candid opinion, which was deliberately elicited from him, was transmitted to the Admiralty's Secretary, John Barrow, by the Royal Society's Secretary, William Thomas Brande, in April 1826, along with the Astronomer Royal's view that there ought not to be two classes of assistants performing the same duties but rewarded in a very unequal manner; to which Barrow replied that the Lords of the Admiralty were sorry that their expectations regarding the acquisition of a superior class of assistants were not going to be realized, and accepted with considerable reluctance the continuance of Thomas Taylor as first assistant.

Pond's subsequent appeal to the Admiralty for increases in the salaries of his second, third, and fourth assistants and for a permanent salary of at least £125 p.a. to be paid to Thomas Ellis and William Rogerson, whom he had just engaged as supernumary helpers, was flatly refused; as was a direct claim for a £50 p.a. increase addressed to Pond from the recognized assistants three years later, which was transmitted by him to the Royal Society Council, from it to the Admiralty, and thence to the Treasury. Latterly, Pond's own ill-health and consequent long-term absences from the Observatory prevented him from keeping a close supervision on his six assistants. Thomas Taylor became an incurable alcoholic, while his family abused Pond's kindness and ineffectual control over his staff until the whole institution degenerated into a state of inefficiency from which it had to be rescued by Pond's hard-headed successor Sir George Airy.

It would, however, be wrong to conclude from this narrow perspective that astronomical science in England was in a state of decline when Pond was harassed into retirement in 1835, although one could be forgiven for thinking so twenty years earlier. The vitalizing force, or elixir, which brought new life into the body of this subject was the Astronomical Society of London, officially founded on 12 January 1820 (or roughly three weeks before Gilbert announced his proposal to establish the Cape Observatory), although the Reverend William Pearson and Francis Baily had been endeavouring throughout the course of the previous decade to promote such a foundation. Pond himself was among its earliest members. A general circumstance attending its inception was expressed by its first president, Sir John Herschel, in the following terms:

> "The end of the eighteenth and beginning of the nineteenth centuries were remarkable for the small amount of scientific movement going on in this country, especially in its more exact departments ... Mathematics were at the last gasp, and Astronomy nearly so—I mean in those members of its frame which depend upon precise measurement and systematic calculation. The chilling torpor of routine had begun to spread itself over all those branches of science which wanted the excitement of experimental research."[9]

This view was undoubtedly conditioned by Herschel's admiration for the sparkling achievements of his own father and his recognition of the serious deficiency in the teaching at Cambridge of analytical methods in geometry and calculus. No one in England had received a training which would have enabled him to make the revolutionary advances in the lunar theory, for example, that had been made over the past fifty years by continental astronomers and mathematicians. Pierre de Laplace's discovery of the cause of the Moon's secular acceleration and other theoretical advances, first incorporated into the *Nautical Almanac* for 1808 (at the same time as Pond was busy translating the former's *Système du Monde*), had the effect of reducing the mean errors in the calculated values of its longitude from 31″ to only 8″; while subsequent computational improvements by Johann Tobias Burg and Johann Karl Burckhardt, whose tables were admittedly constructed principally upon Maskelyne's observations, had brought this figure down even lower (to a mere 5″). The mean error in the predicted latitudes was of the same magnitude. Maximum discrepancies of between four and five times this value were, however, being occasionally detected in 1820. The solar ephemerides in the *Nautical Almanac* from 1813 onwards were based upon new improved solar tables by J. B. Joseph Delambre. Gauss's brilliant method of computing the orbital elements of the increasing number of asteroids now being found was also recognized to transcend all the efforts of his predecessors to develop a satisfactory theory to represent the planetary data in this way; although, as is amply evident from his extensive correspondence on this subject with Wilhelm Olbers and with Maskelyne (from 1802 to 1805), he attached great importance to the 'exquisite' observations with the 'excellent' instruments at Greenwich which the latter had been freely communicating to him in return for his calculated elements for Ceres, Pallas, and Juno.[10]

Herschel's critique should consequently not be construed as an attack against the quality of the observational work being carried out at the Royal Observatory, but rather as a complaint against the low level of skill shown by British astronomers in the construction of theories and in the computational techniques used in reducing the vast quantity of accurate data which Pond and his assistants were in the process of collecting. This accounts to a large extent for the comparative lack of interest shown during the early years of the Astronomical Society's history in following up William Herschel's pioneering researches on nebulae, star clusters, and other objects of astrophysical importance. Instead, men like Francis Baily and Stephen Groombridge, inspired by the example of Bessel (cf. Chapter 5), led them on a campaign of meridian observation, star-corrections, and reform of the *Nautical Almanac*. Attention was also devoted to a variety of other related subjects, including the collation and publication of observations, the education of observers, the measurement of the length of the seconds pendulum for assisting the determination of terrestrial latitude, the improvement of lunar tables as a basis for finding longitude both on land and at sea, the establishment of relations with foreign astronomers, the computation of orbits, the formation of a library, etc. The Royal Society was also deeply involved in many of these investigations.

The zeal with which the Astronomical Society of London pursued its aims throughout the first decade undoubtedly encouraged the Admiralty to believe that the Board of Longitude was rapidly becoming redundant and to arrange for its dissolution in 1828 by the Act 9 George IV, cap. lxvi after it had been seen to have fulfilled the function for which it had been created. However, the clauses relating to the printing, publishing, and vending of the *Nautical Almanac* remained in force, as they still are to-day—over two hundred years after its first appearance under Maskelyne's supervision. Another consequence of the Society's rapidly acquired prestige was William IV's decision on 7 March 1830 to incorporate it under Royal Charter, followed just over six months later by a reconstitution of the Board of Visitors which was to give the Astronomical Society equal control to the Royal Society over the affairs of the Royal Observatory.

The Visitors initially appointed on 9 September 1830 to represent the Royal Society were its new president Davies Gilbert, John Lubbock, Captain Henry Kater, George Peacock, William Pearson, and Richard Sheepshanks. Those representing the Royal Astronomical Society were to be its president Sir James South, Charles Babbage, Francis Baily, Captain Francis Beaufort, Olinthus Gregory, John Herschel, the Savilian Professor of Astronomy at Oxford (S. P. Rigaud), and the Plumian Professor of Astronomy at Cambridge (G. B. Airy). These gentlemen were to receive observations from Greenwich every three months, and it would be left to the discretion of the Lord High Admiral to decide how many copies of these ought to be printed. The first Saturday in June was fixed as the invariable day of their official visitation, and seven of them would form a quorum. Former presidents of both societies would continue to be recognized as Visitors for as long as they wished to attend.

The delay in agreeing the draft of William IV's charter and obtaining his signature in the Society's autograph book as Patron on 15 December 1830 whereupon it at last became entitled to introduce the epithet 'Royal' into its title, resulted in Sir James South no longer being cited as first president—a technicality over which he raised such a storm of protest that he never served on the Council again, although he remained an embarrassment to the Visitors for years to come.

It was the Royal Astronomical Society which took the initiative by asking the President of the Royal Society, Augustus Frederick, Duke of Sussex—the most liberally minded of George III's sons—to call a meeting of the Board of Visitors for the purpose of considering the best means of rapidly publishing the astronomical observations being made both at Greenwich and at the Cape. The newly constituted Board duly met at the Admiralty Office on 19 January 1831, when it was decided to reduce the cost by printing these on paper of inferior quality. Airy, Baily, Herschel and Sheepshanks were appointed to consider the best arrangement of these data. Airy's influence was then apparent in that committee's decision to print in the same form as that used for the Cambridge observations.

Before the end of Pond's term of office, the *Nautical Almanac* had been revised and its production separated from the Royal Observatory, his catalogue of 1113 stars published, and the first accurate public time-signal instituted in the form of the Greenwich time-ball (1833).

We have come, at the conclusion of our story, to the start of a new era in the history of the Royal Observatory. Despite the various shortcomings of that institution and its first six directors which have been indicated in this and the foregoing chapters, it can look back with pride on its early achievements. Not every field of astronomy was cultivated within its walls, but only those of direct benefit to the fostering of astronomical navigation. Cataloguing stellar coordinates was routine work demanding great precision *and* continuity of effort coupled with high-quality instruments and reliable techniques of observation and reduction. Of course, although one generally speaks of finding the positions of stars relative to the equator or ecliptic, what was really being sought was the relative situation of these circles themselves and the points of their intersection; and these are subject to continuous change which must be followed if the motions of the Sun, Moon, and planets are to be found. Baily's comparison of the catalogue of the places of 720 principal stars which Pond appended to the first part of the Greenwich observations for 1829, with those of Bradley and Piazzi reduced to the mean epochs of 1756.00 and 1800.00 respectively, showed how much had still to be done to complete the work which Flamsteed and his successors had pioneered. As he warns his fellow-members of the Royal Astronomical Society on 12 March 1830:

> "—we are at present very far from a knowledge of the true and
> accurate position of most of the fixed stars, including even those
> of very considerable magnitude."[11]

The truth of this statement was borne out by the ultimate major
investigation carried out by Pond in his last two years at Greenwich,
where he attempted to find the zenith distances of a few bright stars
with his newly installed Troughton and Simms 25-foot focus zenith
telescope. He first applied this to the star γ Draconis, but instead of
facing this instrument east, then west, to find the double zenith
distance, he measured the angular distance of that star from another
of the 5th magnitude class having roughly the same zenith distance
to the south as γ Draconis had towards the north and lying just
under half an hour behind it in right ascension. The *sum* of the zenith
distances was easily obtained. By observing the transit of γ Draconis
and immediately afterwards turning the instrument round to observe
the other star on the same (or if clouds intervened, on the following)
night, the *difference* of zenith distance was read off by shifting one of
the wires on the micrometer eyepiece only slightly. Hence both
zenith distances could be evaluated. The results obtained from this
new method were found to be almost identical with those obtained
using the zenith telescope in the conventional manner and independently with the aid of the mural circle, the error never
exceeding $\frac{1}{4}''$. Yet despite this internal agreement, all of these values
were found to be as far as $1''\cdot75$ to the south of the places computed
from Bessel's formula:

zenith distance for the year $1800+t = 2'\ 26''\cdot669 - t.\ 0''\cdot71394 + t^2.\ 0''\cdot001011$

which had been derived from observations covering the sixty-year
period between 1756 and 1816.[12]

The publication by Pond in 1833 of his catalogue of 1113 stars is
generally regarded as his crowning achievement. It embodied several
smaller catalogues, inserted from time to time in the *Nautical
Almanac* and the Greenwich Observations (of which he ultimately
had eight folio volumes printed).[13] His contributions as a whole
were entirely technical, but his reform of the Royal Observatory was
fundamental. During his twenty-four years in office the number of
assistants had been increased from one to six and the results of their

observations were published at quarterly rather than at annual intervals. The double-altitude observations made with the two mural circles from 1825 to 1835 were later reduced by S. C. Chandler for the purposes of his research into the phenomenon of the variation of latitude (alluded to earlier), and he found them to be "a rich mine of stellar measurements" possessing an accuracy that "has been scarcely surpassed anywhere or at any time".[14] His catalogues nevertheless contain slight periodic errors which probably reflect the uncertainty inherent in the system of fundamental stars employed in their construction, rather than any fault in the observational technique or in the quality of the instrumentation. The instruments were among the best of their kind in existence at that time and, as Troughton is said to have remarked: "a new instrument was at all times a better cordial for the astronomer-royal than any which the doctor could supply".[15] Deprived of this medicine by his enforced retirement, Pond did not live long to enjoy his pension of £600 p.a. He died at his residence at Blackheath on 7 September 1836 and his remains were buried in the tomb of Halley in the neighbouring churchyard of Lee.

NOTES

1. R.S. MS. 371, No. 19.
2. *Phil. Trans. R. Soc.*, **114** (1824), pp. 457–470.
3. A copy of this warrant is contained in R.S. MS. 371, No. 64.
4. R.S. Domestic MSS. 2, Nos. 94–111.
5. R.S. Council Minutes of 5 May 1825, taken from R.G.O. MSS. 618 denoted as "Class M, shelf 1, No. 52" (in Airy's arrangement of 1874).
6. R.S. Domestic MSS., **2**, No. 98.
7. *Ibid.*, No. 106.
8. R.S. MSS. Gh. 49.
9. Quoted from Dreyer and Turner, p. 16, but originating from Memoirs of Augustus de Morgan.
10. Full details of this example of international scientific cooperation can be found in Forbes, 1971 a.
11. Baily, 1830, p. 287.
12. The formula is contained in Bessel, 1818, p. 136.
13. These contain a vast quantity of data from 1811 to 1835.
14. D.N.B. 'Pond'; by A.M.C.
15. *Ibid.*

Literary Sources

Manuscripts

THE following list represents only a minute but significant sample of the manuscript correspondence and miscellaneous papers of the first six Astronomers Royal actually examined in the various repositories mentioned, and elsewhere (such as the University libraries of Oxford and Cambridge, the Bibliothèque Nationale, etc.).

BRITISH MUSEUM (B.M.), LONDON

Sloane MS. 871, ff. 2–6. Proposals to King Charles II concerning the longitude.

Birch MS. 4394, ff. 26–108. Original Commission of King Charles II to Lord Brouncker, etc. (1674) containing papers collected by Dr John Pell relating to the committee appointed to examine the longitude schemes of Henry Bond and Le Sieur de St Pierre.

Add. MS. 34712, f. 159. Regulations for the due execution of the office of Astronomer Royal humbly proposed to his Majesty's Consideration.

NATIONAL MARITIME MUSEUM (N.M.M.), GREENWICH

MS. PGR/38/1. Photocopies of the Maskelyne papers in possession of N. Arnold Foster, Salthorp House, Bassettdown, nr. Swindon, Wiltshire, England.

PUBLIC RECORD OFFICE (P.R.O.), LONDON

Admiralty Papers ADM 1 Letters from the Ordnance Office.

Home Office Warrant Books 1, 6.

Board of Ordnance Outletters W.O. 46.

Board of Ordnance Minute Books W.O. 47.

Board of Ordnance Treasurers Ledgers W.O. 48.

Board of Ordnance Miscellaneous Papers W.O. 55.

State Papers (S.P.) Dom. Entry Books 27, 44, 274, 389.

ROYAL GREENWICH OBSERVATORY (R.G.O.), HERSTMONCEUX, SUSSEX

This is the major repository for vast quantities of Flamsteed, Halley and Maskelyne MSS. which still require thorough indexing although they have been broadly classified by the Public Record Office. My own summary of "The Maskelyne Manuscripts at the Royal Greenwich Observatory" has been published in the *Journal for the History of Astronomy*, 5 (1974), 67–69. In addition, there are 57 volumes of Board of Longitude papers covering an enormous range of topics concerning the theory and practice of navigation between 1714 and 1829. My reading has mainly been restricted to five volumes, whose contents are concerned with the Board's actions regarding the recovery of Dr Bradley's astronomical observations from the Clarendon Press (Vol. 4) and the Confirmed Minutes of its meetings from 1737 to 1829 (Vols. 5–8). Precise details of the wealth of untapped material in this section of the R.G.O. archives are to be found in my "Index of the Board of Longitude Papers at the Royal Greenwich Observatory", *loc. cit.*, **1** (1970), 169–179; **2** (1971), 58–70 and 133–145.

ROYAL SOCIETY (R.S.), LONDON

Apart from two volumes of R.S. Council Minutes and the Journal Book, the main items consulted were

MSS. 371 Letters and Documents relating to the Greenwich Observatory.

MSS. 372 Greenwich Observatory Letters and Papers II.

LIX c. 10 MS. Letters Flamsteed to Towneley 1672/3 to 1686/7.

RBC. 5. 73 Peter Perkins' "A Discourse of the Variacon of the Compass" (4 March 1679/80).

Domestic MSS. 2, Nos. 94–111. Stephen Lee's Charges against Greenwich Observations &c. 1824–1827.

STATE-ARCHIVE, HANOVER

Hannover Des. 92 XXXIV No. II, 4, a^1 "Betr. der von Seiten des Prof. Tobias Mayer in Göttingen gelösten englische Preisfrage über die Bestimmung der Longitud. maris 1754–1765".

UNIVERSITY LIBRARY, GÖTTINGEN

Cod. MS. Michaelis 320, pp. 555–569.

This contains further correspondence between Michaelis and the representatives of the Hanoverian government in Hanover and in London, concerning Tobias Mayer's bid for a longitude prize.

Bibliographies

Biographie Universelle, Ancienne et Moderne 85 tomes (Paris, 1811–62).

BIRCH, T. (Ed.). *A General Dictionary, Historical and Critical* (London, 1738).

Note: This is often called Birch's Dictionary or Biographia Britannica.

BRITISH MUSEUM. *General Catalogue of Printed Books.*

CHAMBERS, E. (Ed). *Cyclopaedia: or an universal Dictionary of Arts and Sciences, by E. Chambers; with the supplement and modern improvements incorporated in one alphabet,* by A. Abraham Rees, 5 vols. (London, 1786).

Encyclopaedia Britannica, 7th edition, 21 vols. (Edinburgh, 1842).

GILLISPIE, C. (Ed). *The Dictionary of Scientific Biography,* Vols. 1–7 (in progress).

HOUZEAU, J. and LANCASTER, A. *Bibliographie Générale de L'Astronomie,* 2 tomes (Bruxelles, 1882, 1887).

HUTTON, C. *A Mathematical and Philosophical Dictionary,* 2nd ed., 2 vols. (London, 1815).

Nautical Almanac, 1767–1835.

POGGENDORF, J. (Ed). *Biographisch—Literarisches Handwörterbuch zur Geschichte der exakten Wissenschaften,* 2 Bde. (Leipzig, 1853).

REES, A. (Ed.) *The Cyclopaedia, or Universal dictionary of arts, sciences and literature,* 45 vols. (London, 1819).

Literary Sources

ROYAL SOCIETY. *The Philosophical Transactions* for the years 1665–1835.
   The Record of the Royal Society of London for the promotion of natural knowledge, 4th ed. (London, 1940).

STEPHEN, L. and LEE, S. *The Dictionary of National Biography from the earliest times to 1900*, 22 vols. including Supplement (Oxford, 1908–9).

TAYLOR, E. *The Mathematical Practitioners of Tudor and Stuart England* (Cambridge, 1967).
   The Mathematical Practitioners of Hanoverian England (Cambridge, 1968).

WHITROW, M. (Ed.). *Isis Cumulative Bibliography*, Vols. 1 and 2 (London, 1971).

Books and Articles

The abbreviations used in referring to the periodical literature listed below, which correspond in the case of current publications to those contained in P. Brown and G. B. Stratton's *World List of Scientific Periodicals*, 3 vols., 4th ed. (London, 1963), are as follows:

Ann. Sci.   *Annals of Science.*

Brit. Assoc.   British Association for the Advancement of Science.

Br. J. Hist. Sc.   *British Journal for the History of Science.*

Comm. S. R. S. Gott.   *Commentarii Societatis Regiae Scientiarum Gottingensis.*

Gent. Mag. (n.s.).   *The Gentleman's Magazine and Historical Review* (new series), edited by Sylvanus Urban.

J. Hist. Astr.   *Journal for the History of Astronomy.*

J. Inst. Nav.   *Journal of the Institute of Navigation.*

Mem. astr. Soc.   *Memoirs of the Astronomical Society of London.*

Mem. R. astr. Soc.   *Memoirs of the Royal Astronomical Society.*

Naut. Alm.   *Nautical Almanac.*

Naut. Mag.   *Nautical Magazine.*

Notes Rec. R. Soc. Lond.   *Notes and Records of the Royal Society of London.*

Phil. Mag.   *The London and Edinburgh Philosophical Magazine and Journal of Science.*

Phil. Trans. R. Soc.   *Philosophical Transactions of the Royal Society of London.*

Q. Jl R. astr. Soc.   *Quarterly Journal of the Royal Astronomical Society.*

AIRY, G. "Report on the Progress of Astronomy during the present Century", *Br. Assoc. Rep.*, **2** (London, 1833), 125–189.

AIRY, W. (Ed.). *Autobiography of Sir George Biddell Airy* (Cambridge, 1896).

AITON, E. *The Vortex Theory of Planetary Motion* (London, 1971).

ANON. "John Flamsteed and the Greenwich Observatory", *Gent. Mag.* (n.s.), **1** (1866), 239–252; 378–386; and 549–558.
   "The Greenwich Time Ball", *Engineering*, 7 November 1952.
   An Account of the Proceedings, in order to the Discovery of Longitude at Sea; relating principally to the Time-Piece of Mr. John Harrison, etc., 2nd ed. (London, 1763).
   A Narrative of the Proceedings relative to the discovery of the Longitude at Sea; by Mr. John Harrison's Time-Keeper; subsequent to those published in the Year 1763 (London, 1765).

ARMITAGE, A. "The Astronomical Work of Nicolas-Louis de la Caille", *Ann. Sci.*, **12** (1956), 163–191.
Edmond Halley (London, 1966).

AUWERS, A. *Neue Reduktion der Bradleyschen Beobachtungen*, 3 Bde. (St Petersburg, 1882–1903).
Bearbeitung der Bradleyschen Beobachtungen an den alten Meridianinstrumenten der Greenwicher Sternwarte, 3 Bde. (Leipzig, 1912–14).

BAILLY, J. *Histoire de l'astronomie moderne*, 3 tomes (Paris, 1778–82).

BAILY, F. *New Tables for facilitating the computation of Precession, Aberration and Nutation of 2881 principal fixed Stars: together with a Catalogue of the same reduced to January 1, 1830* (London, 1827).
"A Catalogue of the Positions (in 1690) of 564 Stars observed by Flamsteed, but not inserted in his British Catalogue; together with some Remarks on Flamsteed's Observations", *Mem. astr. Soc.*, **4** (1831), 129–164.
"On Mr. Pond's recent Catalogue of the Places of 720 principal Stars, compared with the Places of the same Star in the Catalogue of the Society: with remarks on the Differences between the two Catalogues", *ibid.*, 255–292.
An Account of the Revd John Flamsteed, the First Astronomer-Royal; etc. To which is added his British Catalogue of Stars, corrected and enlarged (London, 1835).
"Some Account of the Astronomical Observations made by Dr. Edmund Halley, at the Royal Observatory at Greenwich", *Mem. R. astr. Soc.*, **8** (1835), 169–190.
The Catalogue of Stars of the British Association for the Advancement of Science (London, 1845).

BENTHEIM, H. *Engländischer Kirch- und Schulen-Statt* (Lüneburg, 1694).

BERNOULLI, J. *Lettres Astronomiques* (Berlin, 1771). Cf. pp. 65–100.

BERTRAND, J. *L'Academie des Sciences et les Academiciens de 1666 a 1793* (Paris, 1869).

BESSEL, F. *Fundamenta Astronomiae pro anno MDCCLV, deducta ex observationibus . . . J. Bradley in Specula astronomica Grenovicensi . . . institutis, etc.* ([Königsberg], 1818).
Tabulae Regiomontanae Reductionum Observationum Astronomicarum ab anno 1750 usque ad annum 1850 computatae (Regiomonti Prussorum, 1830).
Briefwechsel zwischen W. Olbers und F. W. Bessel / herausgegeben von Adolph Erman, 2 vols. (Leipzig, 1852).

BEVIS, J. (Ed.). See HALLEY (1749).

BIGOURDAN, G. *Les Premières Sociétés Savantes de Paris et les Origines de l'Academie des Sciences* (Paris, 1919).

BIRD, J. *The Method of Dividing Astronomical Instruments* (London, 1767).
The Method of Constructing Mural Quadrants (London, 1768).

BLACKWELL, D. "The Discovery of Stellar Aberration", *Q. Jl R. astr. Soc.*, **4** (1963), 44–46.

Literary Sources

BLISS, N. a. "Observations on the Transit of Venus over the Sun, on the 6th of June 1761: In a Letter to the Right Honourable George Earl of Macclesfield, President of the Royal Society", *Phil. Trans. R. Soc.*, **52** (1761), 173–177. b. "A second Letter to the Right Hon. the Earl of Macclesfield, President of the Royal Society, concerning the Transit of Venus over the Sun, on the 6th of June 1761", *ibid.*, 232–250.
"Observations on the Eclipse of the Sun, April 1, 1764: In a Letter to the Right Honourable James Earl of Morton, Pres. R.S.", *Phil. Trans. R. Soc.*, **54** (1764), 141–144.

BRADLEY, J. "A Letter from the Reverend Mr. James Bradley Savilian Professor of Astronomy at Oxford, and F.R.S. to Dr. Edmond Halley Astronom. Reg. &c. giving an Account of a new discovered Motion of the Fix'd Stars", *Phil. Trans. R. Soc.*, **35** (1728), 637–661.
"A Letter to the Right honourable George Earl of Macclesfield concerning an apparent Motion observed in some of the fixed Stars", *Phil. Trans. R. Soc.*, **45** (1748), 1–43.
"Observations upon the Comet that appeared in the Months of September and October 1757, made at the Royal Observatory by Ja. Bradley, etc.". *Phil. Trans. R. Soc.*, **50** (1757), 408–415.
Astronomical Observations made ... from the year MDCCL, to the year MDCCLXII, etc. Vol. 1 (Oxford, 1798); Vol. 2 (Oxford, 1805).
Miscellaneous Works and Correspondence (Oxford, 1832), edited by S. P. Rigaud.

BREWSTER, D. *Memoirs of the Life, Writings, and Discoveries of Sir Isaac Newton,* 2 vols. (Edinburgh, 1860).

BRINKLEY, J. "An Account of Observations made at the Observatory of Trinity College, Dublin, with an Astronomical Circle, eight feet in diameter, which appear to point out an annual parallax in certain fixed stars", etc. *Trans. R. Irish Acad.*, **12** (1815), 33–75.
"On the parallax of certain fixed stars", *Phil. Trans. R. Soc.*, **108** (1818), 275–302.
"On the north polar distances of the fixed stars", *Phil. Trans. R. Soc.*, **114** (1824), 50–84.

BRUHNS, C. *Die Astronomische Strahlenbrechung in ihrer historischen Entwicklung* (Leipzig, 1861).

BRUNET, P. *La Vie et l'Ouvre de Clairaut, 1713–1765* (Paris, 1952).

BUSCH, A. *Reduction of the Observations made by Bradley at Kew and Wansted, to determine the quantities of Aberration and Nutation* (Oxford, 1838).

CASSINI, J. D. *Les Élémens de l'astronomie vérifiés* (Paris, 1684).

CHAPIN, S. "Expeditions of the French Academy of Sciences, 1735", *Navigation,* **3** (1951), 120–122.

COHEN, I. *Roemer and the First Determination of the Velocity of Light* (New York, 1944).

COLLINS, J. *The Sector on a Quadrant, etc.* (London, 1658).

COMMONS JOURNAL, **17** (11 June 1714), 677–678.

COTTER, C. *A History of Nautical Astronomy* (London, 1968).

CUDWORTH, W. *Life and Correspondence of Abraham Sharp, The Yorkshire Mathematician and Astronomer, and Assistant of Flamsteed; with Memorials of his Family and Associated Families* (London, 1889).

DELAMBRE, J. *Histoire de l'Astronomie moderne*, 2 tomes (Paris, 1821).
Histoire de l'Astronomie au Dix-Huitième Siècle (Paris, 1827).

DOLLOND, P. "A Letter from Mr. Peter Dollond, to Nevil Maskelyne, F.R.S. and Astronomer Royal; describing some Additions and Alterations made to Hadley's Quadrant, to render it more serviceable at Sea", *Phil. Trans. R. Soc.*, **62** (1772), 95–98.

DREYER, J. "Flamsteed's Letters to Richard Towneley", *Observatory*, **45** (1922), 280–294.

DREYER, J. and TURNER, H. *History of the Royal Astronomical Society 1820–1920* (London, 1923).

EARNSHAW, T. *Longitude. An Appeal to the Public, etc.* (London, 1808).

EDINBURGH REVIEW, **62** (1836), 359–397—review of BAILY (1835 a).

EGGEN, O. "Flamsteed and Halley", *Occasional Notes R. astr. Soc.*, **3** (1958), 210–221.

ERMAN, A. See BESSEL (1852).

EVANS, J. "The Royal Observatory", *Juvenile Tourist* (1810), 332–338.

FAIVRE, J.-P. "Savants et Navigateurs: Un Aspect de la Coopération Internationale entre 1750 et 1840", *Cahiers d'Histoire Mondiale*, **10** (1966), 98–124.

FAURÉ-FREMIET, E. "Les Origines de l'Académie des Sciences de Paris", *Notes Records R. Soc. Lond.*, **20** (1966), 20–31.

FLAMSTEED, J. "De Temporis Aequatione Diatriba. Numeri ad Lunae Theoriae Horrocciae". See HORROX (1673).
"The Doctrine of the Sphere, grounded on the motion of the earth and the ancient Pythagorean or Copernican system of the World". See MOORE (1681).
A Correct Tide-Table, shewing the true times of the high-waters at London-Bridge to every day in the year 1688 (London, [1687]).
Historiae coelestis libri duo quorum prior exhibet catalogum stellarum fixarum Britannicum. etc. (Londini, 1712), edited by Edmond Halley.
Historia Coelestis Britannica, 3 vols. (Londini, 1725).

FLAMSTEED, MARGARET and HODGSON, J. *Atlas Coelestis* (London, 1729).

FORBES, E. "The Foundation and Early Development of the Nautical Almanac", *J. Inst. Nav.*, **18** (1965), 391–401.
a. "The Origin and Development of the Marine Chronometer", *Ann. Sci.*, **22** (1966), 1–25.
b. "Tobias Mayer's Lunar Tables", *ibid.*, 105–116.
"The Bicentenary of the Nautical Almanac (1767)", *Br. J. Hist. Sci.*, **3** (1967), 393–394.
"Dr. Bradley's Astronomical Observations", *Q. Jl R. astr. Soc.*, **6** (1965) 321–328.
"Tobias Mayer (1723–1762): a case of forgotten genius", *Br. J. Hist. Sci.*, **5** (1970), 1–20.

a. "Who discovered Longitude at Sea?", *Sky Telescope*, **39** (1971), 3–6.

b. "The Correspondence between Carl Friedrich Gauss and the Rev. Nevil Maskelyne (1802–5)", *Ann. Sci.*, **27** (1971), 213–237.

c. "The Rational Basis of Kepler's Laws", *J. Br. astr. Assoc.*, **81** (1971), 33–36.

d. *The Euler-Mayer Correspondence (1751–1755); a new perspective on eighteenth-century developments in the lunar theory* (London, 1971).

e. *Tobias Mayer's Opera Inedita* (London, 1971).

The Birth of Navigational Science (H.M.S.O., 1974).

The Gresham Lectures of John Flamsteed, F.R.S., First Astronomer Royal 1681–84 (Mansell, 1975).

GOULD, R. T. *The Marine Chronometer: its History and Development* (London, 1923); 2nd ed. (London, 1960).

GRANT, R. *History of Physical Astronomy* (London, 1852).

GREEN, C. and COOK, J. "Observations made, by appointment of the Royal Society, at King George's Island in the South Sea", *Phil. Trans. R. Soc.*, **61** (1771), 397–421.

GREGORY, D. *The Elements of Physical and Geometrical Astronomy*, Vol. 2 (London, 1726).

GUYOT, E. *Histoire de la détermination des longitudes* (La Chaux-de-Fonds, 1955).

HALL, A. "Robert Hooke and Horology", *Notes Records R. Soc. Lond.*, **8** (1951), 167–177.

HADLEY, J. "The description of a new instrument for taking angles", *Phil. Trans. R. Soc.*, **37** (1731), 147–157.

A description of a new instrument for taking the latitude or other altitudes at sea (London, 1734).

HALLEY, E. *Catalogus Stellarum Australium* (Londini, 1679).

"A theory of the Variation of the magnetical Compass", *Phil. Trans. R. Soc.*, **13** (1683), 208–221.

"An Account of the cause of the change in the Variation of the Magnetical Needle; with an hypothesis of the structure of the internal parts of the Earth", *Phil. Trans. R. Soc.*, **17** (1692), 563–578.

A New and Correct Sea Chart of the Whole World shewing the Variations of the Compass as they were found in the year M.D.C.C. (London, 1702).

(Ed). See FLAMSTEED (1712).

"A Series of observations on the Planets, chiefly of the Moon, made near London ... being a proposal how to find the Longitude &c". See STREETE (1716).

"A Proposal of a method for finding the longitude at sea within a degree, or twenty leagues. With an account of the progress he hath made therein, by a continued series of accurate observations of the Moon, taken by himself at the Royal Observatory at Greenwich", *Phil. Trans. R. Soc.*, **37** (1731–2), 185–195.

Edmundi Halleii Astronomi dum viveret Regii Tabulae Astronomicae, accedunt de usu Tabularum Praecepta (Londini, 1749), edited by John Bevis.
Astronomical Tables with Precepts both in English and Latin for computing the places of the Sun, Moon, &c. (London, 1752).

HARRISON, J. *Remarks on a Pamphlet lately published by the Rev. Mr. Maskelyne, under the Authority of the Board of Longitude* (London, 1767).

HARTLEY, H. (Ed.). *The Royal Society: its Origins and Founders* (London, 1960).

HECKER, J. *Motuum Caelestium Ephemerides ab anno 1660 and 1680 ex observationibus correctis nobilissimorum Tychonis Brahei, & Ioh. Kepleri hypothesibus physicis, Tabulisque Rudolphinis, ad meridianum Uraniburgicum in freto Cymbrico. Cum Introductione in eas* (Parisiis, 1666).

HELLMAN, C. DORIS. "John Bird (1709–1776) Mathematical Instrument-Maker in the Strand", *Isis*, **17** (1932), 127–151.
"George Graham—Maker of Horological and Astronomical Instruments", *Vassar Journal of Undergraduate Studies*, **5** (1931), 221–251.

HETHERINGTON, N. "The first measurements of stellar parallax", *Ann. Sci.*, **28** (1972), 319–325.

HEVELIUS, J. *Mercurius in Sole visus Gedani ... 1661 ... 3 Maji st. n ...; cui annexa est, Venus in Sole pariter visa, Anno 1639, d. 24 Nov. st. v. Liverpoliae, a Jeremia Horroxio: etc.* (Gedani, 1662).
Machina Caelestis pars prior (Gedani, 1673).
Machina Caelestis pars posterior (Gedani, 1679).
Annus Climactericus (Gedani, 1685).

HOLLIS, H. "The Greenwich Assistants during 250 Years", *Observatory*, **48** (1925), 388–398.

HOOKE, R. *Animadversions on the First Part of the Machina Coelestis* (London, 1674).

HORNSBY, T. (Ed.). See BRADLEY (1798).

HORROX, J. *Opera Postuma; viz. Astronomia Kepleriana, defensa & promota: Excerpta ex Epistolis ad Crabtraeum suum. Observationum Coelestium Catalogus. Lunae Theoria nova. Accedunt Guilielmi Crabtraei, Mancestriencis, Observationes Coelestes. In calce adjiciuntur Johannis Flamstedii, Derbiensis, De Temporis Aequatione Diatriba. Numeri ad Lunae Theoriam Horroccianam* (Londini, 1673), edited by John Wallis.

HUMBERT, P. *L'Astronomie en France au Dix-Septième Siècle* (Paris, 1952).

JOURNAL OF THE HOUSE OF COMMONS, **17** (25 May–8 July, 1714), 641–2, 671, 677–8, 686–7, 692, 711–2, 715–6, 721.

KIRK, R. *Mr. Pepys upon the State of Christ-Hospital* (Philadelphia, 1935).

LACAILLE, ABBÉ DE. "Extract of a Letter from the Abbé De la Caille, of the Royal Academy of Sciences at Paris, and F.R.S. to William Watson M.D.F.R.S., recommending to the Rev. Mr. Nevil Maskelyne F.R.S. to make at St. Helena a series of Observations for discovering the Parallax of the Moon", *Phil. Trans. R. Soc.*, **52** (1761), 17–21.

LALANDE, J. *Astronomie*, 2nd ed., 4 tomes (Paris, 1771–81).

LAPLACE, P. DE. "Sur l'équation séculaire de la lune", *Hist. Acad. R. Sci. Paris*, Année 1786 (1788), 235–264.
The System of the World . . . Translated . . . by J. Pond (London, 1809).

LAURIE, P. a. "The Board of Visitors of the Royal Observatory—I: 1710–1830", *Q. Jl R. astr. Soc.*, **7** (1967), 169–185.
b. "The Board of Visitors of the Royal Observatory—II: 1830–1965", *ibid.*, 334–353.

LAURIE, P. and WATERS, D. "James Bradley's New Observatory and Instruments", *Q. Jl R. astr. Soc.*, **4** (1963), 55–61.

MACPIKE, E. *Correspondence and Papers of Edmond Halley* (Oxford, 1932).
Hevelius, Flamsteed and Halley: three contemporary astronomers and their mutual relations (London, 1937).

McCREA, W. "James Bradley 1693–1762", *Q. Jl R. astr. Soc.*, **4** (1963), 38–40.
"The Significance of the Discovery of Aberration", *ibid.*, 41–43.

McKIE, D. "The Origins and Foundation of the Royal Society of London", in HARTLEY (Ed.), 1–37.

MAIRE, C. and BOSCOVICH, R. J. *De Litteraria Expeditione per Pontificiam Ditlonem ad Dimetiendos duos Meridiani Gradus et corrigendam Mappam Geographicam jussu, et auspiciis Benedicti XIV. Pont Max. suscepta a patribus Societ. Jesu Christophoro Maire et Rogerio Josepho Boscovich* (Romae, 1755).

MARGUET, F. *Histoire de la Longitude à la Mer au XVIIIè Siècle en France* (Paris, 1917).

MASKELYNE, N. "A Proposal for discovering the Annual Parallax of Sirius", *Phil. Trans. R. Soc.*, **51** (1760), 889–895.
a. "A Letter from the Rev. Nevil Maskelyne, A.M. Fellow of Trinity College, in the University of Cambridge, and F.R.S. to the Rev. Dr. Birch, Secretary to the Royal Society; containing a Theorem of the Aberration of the Rays of Light refracted through a Lens, on account of the imperfection of the spherical Figure", *Phil. Trans. R. Soc.*, **52** (1761), 17–21.
b. "A Letter from the Rev. Nevil Maskelyne, M.A. F.R.S., to William Watson, M.D. F.R.S.", *ibid.*, 26–28.
c. "An Account of the Observations made on the Transit of Venus, June 6, 1761, in the Island of St. Helena: In a Letter to the Right Honourable George Earl of Macclesfield, President of the Royal Society, from the Rev. Nevil Maskelyne M.A. and F.R.S.", *ibid.*, 196–201.
d. "Observations on a Clock of Mr. John Shelton, made at St. Helena: In a Letter to the Right Honourable Lord Charles Cavendish, Vice-President of the Royal Society", *ibid.*, 434–443.
e. "A Letter from the Rev. Nevil Maskelyne, M.A. F.R.S. to the Rev. Thomas Birch, D.D. Secretary to the Royal Society: containing the Results of Observations of the Distance of the Moon from the Sun and fixed Stars, made in a Voyage from England to the Island of St. Helena, in order to determine the Longitude of the Ship, from Time to Time; together with the whole Process of Computation used on this Occasion", *ibid.*, 558–577.

f. "Observations on the Tides in the Island of St. Helena: etc." *ibid.*, 586–606.

The British Mariner's Guide (London, 1763).
a. "Concise Rules for computing the Effects of Refraction and Parallax in varying the Apparent Distance of the Moon from the Sun or a Star; also an easy Rule of Approximation for computing the Distance of the Moon from a Star, the Longitudes and Latitudes of both being given, with Demonstrations of the same", *Phil. Trans. R. Soc.*, **54** (1764), 263–276.

b. "Some Remarks upon the Equation of Time, and the true Manner of computing it", *Ibid.*, 336–347.

c. "Astronomical Observations made at the Island of St. Helena", *ibid.*, 348–386.
(Ed.). *The Nautical Almanac* (from 1767 to 1811).
Tables requisite to be used with the Nautical Ephemeris for finding the Latitude and Longitude at Sea (London, 1767), 2nd ed. (London, 1781).
See MAYER (1767).
An Account of the going of Mr. John Harrison's Watch at the Royal Observatory, from May 6th 1766 to March 4th 1767. Together with the Original Observations and Calculations of the same (London, 1767).

a. "Introduction to two Papers of Mr. John Smeaton, F.R.S.", *Phil. Trans. R. Soc.*, **58** (1768), 152–153.

b. "Eclipses of the three first Satellites of Jupiter, observed, at the Royal Observatory at Greenwich, in the years 1762, 1763 and 1764, during the time that the late Rev. Nathaniel Bliss, A.M. F.R.S. Professor of Geometry in the University of Oxford was Astronomer Royal; etc.", *ibid.*, 201–202.

c. "Introduction to the following Observations, made by Messieurs Charles Mason and Jeremiah Dixon, for determining the Length of a Degree of Latitude, in the Provinces of Maryland and Pennsylvania, in North America", *ibid.*, 270–273.

d. "The Length of a Degree of Latitude in the Province of Maryland and Pennsylvania, deduced from the foregoing Operations", *ibid.*, 323–328.

e. "Observations of the Transit of Venus over the Sun, and the Eclipse of the Sun, on June 3, 1769; made at the Royal Observatory", *ibid.*, 355–365.
See MAYER (1770).
"Description of a Method of measuring Differences of Right Ascension and Declination, with Dollond's Micrometer, together with other new Applications of the same", *ibid.*, **61** (1771), 536–546.

a. "Directions for using the common Micrometer, taken from a Paper in the late Dr. Bradley's Hand-writing", *ibid.*, **62** (1772), 46–53.

b. "Remarks on the Hadley's Quadrant, tending principally to remove the Difficulties which have hitherto attended the Use of the Back-observation, and to obviate the Errors that might arise from a Want of Parallelism in the two Surfaces of the Index-Glass", *ibid.*, 99–122.

a. "A Proposal for measuring the Attraction of some Hill in this Kingdom by Astronomical Observations", *ibid.*, **65** (1775), 495–499.

b. "An Account of Observations made on the Mountain Schehallien for finding its Attraction", *ibid.*, 500–542.

Astronomical Observations made at the Royal Observatory at Greenwich from the year 1765 to the year 1774 (London, 1776).

"Account of a new Instrument for measuring small Angles, called the prismatic Micrometer", *Phil. Trans. R. Soc.*, **67** (1777), 799–813.

"Advertisement of the expected Return of the Comet of 1532 and 1661 in the Year 1788", *ibid.*, **76** (1786), 426–431.

a. "Concerning the Latitude and Longitude of the Royal Observatory at Greenwich; with Remarks on a Memorial of the late M. Cassini de Thury", *ibid.*, **77** (1787), 151–187.

b. *Astronomical Observations made at the Royal Observatory at Greenwich from the year 1775 to the year 1786*, **2** (London, 1787).

An Answer to a pamphlet entitled 'A Narrative of Facts', lately published by Mr. Thomas Mudge, Junior, relating to some Time-Keepers constructed by his Father Mr. Thomas Mudge, etc. (London, 1792).

"Observations of the Comet of 1793, made by the Rev. Nevil Maskelyne, D.D. F.R.S. Astronomer Royal and other Observers", *Phil. Trans. R. Soc.*, **83** (1793), 55.

"An Account of an Appearance of Light, like a Star, seen lately in the dark Part of the Moon, by Thomas Stretton, in St. John's Square, Clerkenwell, London; with Remarks upon this Observation, and Mr. Wilkins's", *ibid.*, **84** (1974), 435–440.

Astronomical Observations made at the Royal Observatory at Greenwich from the year 1787 to the year 1798, **3** (London, 1799).

"On a new Property of the Tangents of three Arches trisecting the Circumference of a Circle", *Phil. Trans. R. Soc.*, **98** (1808), 122–123.

Astronomical Observations made at the Royal Observatory at Greenwich from the year 1799 to the year 1810, **4** (London, 1811).

MASON, C. *Mayer's Lunar Tables, improved by Mr. Charles Mason* (London, 1787).

MAUNDER, E. *The Royal Observatory, Greenwich; a glance at its history and work* (London, 1900).

MAY, W. "Early reflecting instruments", *Naut. Mag.*, **145** (1945), 21–26.
"The Compass Makers of Deptford Dockyard", *ibid.*, **163** (1950), 386–390.
"Navigational Accuracy in the Eighteenth Century", *J. Inst. Nav.*, **6** (1953), 71–73.
"Naval Compasses in 1707", *ibid.*, 405–409.

MAYER, T. "Novae tabulae motuum solis et lunae", *Comm. S. R. S. Gott.*, **2** (1753), 383–430.
"Tabularium lunarium in Commentt. S.R. Tom. II contentarum usus in investiganda longitudine maris", *Comm. S. R. S. Gott.*, **3** (1754), 375–384.
Theoria lunae juxta systema Newtonianum (Londini, 1767), edited by Nevil Maskelyne.

Tabulae motuum solis et lunae novae et correctae auctore Tob. Mayer: quibus accedit methodus longitudinum promota eodem auctore (Londini, 1770), edited both in Latin and in English by Nevil Maskelyne.

MERCER, R. *John Arnold and Son, Chronometer Makers 1762–1843* (London, 1972).

MOODY, A. "Early Units of Measurement and the Nautical Mile", *J. Inst. Nav.*, 5 (1952), 262–270.

MOORE, J. *A New Systeme of the Mathematicks* (London, 1681).

MOXON, J. See WRIGHT (1657).

MUDGE, T. *A Narrative of Facts relating to some Time-Keepers, constructed by Mr. Thomas Mudge for the discovery of Longitude at sea: together with Observations upon the conduct of the Astronomer Royal respecting them* (London, 1792).
A Reply to the Answer of the Rev. Dr. Maskelyne Astronomer Royal to A Narrative of Facts, etc. (London, 1792).
A Description, with plates, of the Time-Keeper invented by the late Mr. Thomas Mudge, etc. (London, 1799).

MURRAY, C. "The Royal Observatory and its Work", *British Almanac and Companion for 1885*, pp. 115–133.

NEPTUNE (pseudonym), "The Time-Ball at Greenwich", *Naut. Mag.*, 4 (1835), 584–586.

NIEBUHR, C. "Über die Löngenbestimmung im Orient u.s.w.", in F. X. von Zach (Ed.) *Monatliche Correspondenz zur Beförderung der Erd- und Himmels- Kunde*, 4 (1801), 240–253.

OLMSTED, J. "The Scientific Expedition of Jean Richer to Cayenne", *Isis*, 34 (1942), 117–128.

OLOFF, J. *Excerpta ex Literis . . . Dn. Johannem Hevelium . . . de Rebus Astronomicis . . . scriptis* (Gedani, 1683).

PARLIAMENTARY DEBATES 1713–1714.

PEARSON, W. *Introduction to Practical Astronomy*, 2 vols. (London, 1829).

PINGRÉ, G. "A Supplement to Mons. Pingré's Memoir on the Parallax of the Sun: In a Letter from him to the Royal Society, Translated by M. Maty, M.D. F.R.S.", *Phil. Trans. R. Soc.*, 54 (1764), 152–160.

POND, J. "On the Declinations of some of the principal fixed Stars; with a Description of an Astronomical Circle, and some Remarks on the Construction of Circular Instruments", *ibid.*, 96 (1806), 420–454.
(transl.)   See LAPLACE (1809).
a. "Observation of the Summer Solstice, 1812, at the Royal Observatory", *ibid.*, 103 (1813), 27–30.
b. "A Catalogue of North Polar Distances of some of the principal fixed Stars", *ibid.*, 75–76.
c. "Observation of the Winter Solstice of 1812, with the Mural Circle at Greenwich", *ibid.*, 123–125.

d. "Catalogue of North Polar Distances of Eighty-four principal fixed Stars, deduced from Observations made with the Mural Circle at the Royal Observatory", *ibid.*, 280–304.

"Determination of the North Polar Distances and proper motion of thirty fixed Stars", *ibid.*, **105** (1815), 384–388.

Astronomical Observations made at the Royal Observatory at Greenwich, in the Years M.DCCC.XI. M.DCCC.XII. and M.DCCC.XIII, 8 vols. (London, 1815–1835).

a. "On the parallax of the fixed stars", *Phil. Trans. R. Soc.* **107** (1817), 158–175.

b. "On the parallax of the fixed stars", *ibid.*, 353–362.

a. "On the different methods of constructing a catalogue of fixed stars", *ibid.*, **108** (1818), 405–416.

b. "On the parallax of α Aquilae", *ibid.*, 477–480.

c. "On the parallax of the fixed stars in right ascension", *ibid.*, 481–485.

"A Letter from John Pond, Esq. Astronomer Royal, to Sir Humphry Davy, Bart. President of the Royal Society, relative to a derangement in the Mural Circle at the Royal Observatory", *ibid.*, **112** (1822), 86–88.

a. "On the changes which have taken place in the declination of some of the principal fixed Stars", *ibid.*, **113** (1823), 34–52.

b. "On the parallax of α Lyrae", *ibid.*, 53–72.

c. "On certain changes which appear to have taken place in the positions of some of the principal fixed Stars", *ibid.*, 529–540.

"On the annual variations of some of the Principal fixed Stars", *ibid.*, **115** (1825), 510–512.

a. "On the Latitude of the Royal Observatory at Greenwich", *Mem. astr. Soc.*, **2** (1826), 317–319.

b. "Supplement to a former Paper 'On the Latitude of the Royal Observatory of Greenwich' ", *ibid.*, 529–530.

"On the Obliquity of the Ecliptic, as observed by Bradley, compared with later Observations", *Mem. R. astr. Soc.*, **5** (1833), 9–12.

"Some Suggestions relative to the best Method of employing the New Zenith Telescope lately erected at the Royal Observatory", *Phil. Trans. R. Soc.*, **124** (1834), 209–212.

"Continuation of a former Paper on the Twenty-five Feet Zenith Telescope lately erected at the Royal Observatory", *ibid.*, **125** (1835), 145–151.

PURVER, M. *The Royal Society: Concept and Creation* (London, 1967).

QUARRELL, W. and MARE, M. (Eds.). *London in 1710. From the travels of Zacharias Conrad von Uffenbach* (London, 1934).

QUILL, H. *John Harrison—The Man Who found Longitude* (New York, 1966).

RAMSDEN, J. *Description of an Engine for Dividing Mathematical Instruments* (London, 1777).
Description of an Engine for Dividing Strait Lines on Mathematical Instruments (London, 1779).

RICHER, J. *Mémoires de l'Academie Royale des Sciences de Paris depuis 1666 jusqu' à 1699*, **7** (1729), 233–326.

RIGAUD, S. (Ed.). See BRADLEY, J. (1832).
"Observations on a Note respecting Mr. Whewell, which is appended to No. CX of the Quarterly Review", *Phil. Mag.*, **8** (1836), 218–225.
"Some Particulars respecting the principal Instruments at the Royal Observatory at Greenwich, in the time of Dr. Halley", *Mem. R. astr. Soc.*, **9** (1836), 205–227.
Correspondence of Scientific Men of the seventeenth century, **2** (Oxford, 1841).

ROBERTSON, A. (Ed.). See BRADLEY (1805).

ROBINSON, H. W. and ADAMS, W. (Eds.). *The diary of Robert Hooke, M.A. M.D., F.R.S. 1672–1680* (London, 1935).

RONAN, C. Edmond Halley: genius in eclipse (London, 1969).

ROY, W. "An Account of the Mode proposed to be followed in determining the relative Situation of the Royal Observatories of Greenwich and Paris", *Phil. Trans. R. Soc.*, **77** (1787), 188–226.

S., C. "On Whiston, Halley, and the Quarterly Reviewer of the 'Account of Flamsteed' ", *Phil. Mag.*, **8** (1836), 225–226.

S—s, "The Royal Observatory, Greenwich", *The Weekly Visitor*, cxxi (1835), 1–2; cxxvi (1835), 41–42; cxxvii (1835), 53–55; cxxviii (1835), 63–64; cxxix (1835), 67–69.

SADLER, D. "The Bicentenary of the Nautical Almanac", *J. Inst. Nav.*, **21** (1968), 6–18.

SARTON, G. "Discovery of the Aberration of Light", *Isis*, **16** (1931), 233–265.
"Discovery of the main nutation of the earth's axis", *ibid.*, **17** (1932), 333–383.

SCOTT, J. (Ed.). *The Correspondence of Isaac Newton*, **4** (London, 1967).

SHEEHAN, J. "The watchmaker and the scientist—an almost forgotten controversy", *U.S. Naval Inst. Proc.*, **70** (1944), 161–166.

SKEMPTON, A. and BROWN, JOYCE. "John and Edward Troughton, Mathematical Instrument Makers", *Notes Rec. R. Soc. Lond.*, **27** (1973), 233–262.

SMEATON, J. "Observations on the Graduation of Astronomical Instruments; with an Explanation of the Method invented by the late Mr. Henry Hindley, of York, Clock-maker, to divide Circles into any given Number of Parts", *Phil. Trans. R. Soc.*, **76** (1786), 1–47.

SMITH, R. *A Compleat System of Opticks* (Cambridge, 1788).

SOLVER, C. and MARCUS, G. "Dead Reckoning and the Ocean Voyages of the Past", *Mariners Mirror*, **44** (1958), 18–34.

SPENCER JONES, H. *The Royal Observatory, Greenwich* (London, 1946).
"The Development of Navigation", *J. Inst. Nav.*, **1** (1948), 1–12.

STEWART, A. "The Discovery of Stellar Aberration", *Scient. Am.*, **210** (1964), 100–108.

STREET[E], T. *Astronomia Carolina . . . to which is added a series of observations on the planets, chiefly of the Moon, made near London: with a sextant of near six foot radius; in order to find out the Lunar Theory a Posteriori. Being a Proposal how to find the Longitude &c. By Dr. Edmund Halley*, 3rd ed., corrected (London, 1716).

TAYLOR, E. "Old Henry Bond and the Longitude", *Mariners Mirror*, **25** (1939), 162–169.

"The Dawn of Modern Navigation", *J. Inst. Nav.*, **1** (1948), 283–289.

"Navigation in the Days of Captain Cook", *ibid.*, **21** (1968), 256–276.

THOREN, V. "New Light on Tycho's Instruments", *J. Hist. Astr.*, **4** (1973), 25–45.

THROWER, N. "The Discovery of Longitude Observations on Carrying Time-keepers for Determining longitude at Sea, 1530–1770", *Navigation*, **5** (1953), 375–381.

TROLLOPE, W. *A History of the Royal Foundation of Christ's Hospital* (London, 1834).

TURNBULL, H. (Ed.). *The Correspondence of Isaac Newton*, 3 vols. (London, 1959–61).

WALLIS, J. (Ed.). See HORROX (1673).

Opera Mathematica, **3** (Oxonii, 1699).

WARD, J. *The Lives of the Professors of Gresham College* (London, 1740).

WARGENTIN, P. "A Letter from Mr. Peter Wargentin, F.R.S. Secretary to the Royal Academy of Sciences at Stockholm, to the Rev. Nevil Maskelyne, B.D., F.R.S. and Astronomer Royal; concerning the Difference of Longitude of the Royal Observatories at Paris and Greenwich, resulting from the Eclipses of Jupiter's first Satellites, observed during the last Ten Years: to which is added, a Comparative Tables of the corresponding Observations of the First Satellite, made in the principal Observatories", *Phil. Trans. R. Soc.*, **67** (1777), 162–186.

WHATTON, A. *Memoir of the Life and Labors of the Rev. Jeremiah Horrox, etc.* (London, 1859).

WHEWELL, W. a. "Newton and Flamsteed, etc." *Phil. Mag.*, **8** (1836), 139–147.
b. "Remarks on a Note on a Pamphlet entitled 'Newton and Flamsteed' in No. cx of the Quarterly Review", *ibid.*, 211–215.
c. "To the Editor of the Cambridge Chronicle", *ibid.*, 215–218.

WHISTON, W. and DITTON, H. *New Method of Discovering the Longitude* (London, 1714).

WING, V. *Astronomia Britannica* (Londini, 1669).

WOLF, C. *Histoire de l'Observatoire de Paris de sa Fondation à 1793* (Paris, 1902).

WOOLF, H. *The Transits of Venus. A Study of Eighteenth-Century Science* (Princeton, 1969).

WOOLLEY, R. "James Bradley, Third Astronomer Royal", *Q. Jl R. astr. Soc.*, **4** (1963), 47–52.

WRIGHT, E. *Certain Errors in Navigation . . . detected and corrected. With many additions that were not in the former Editions* (London, 1657), edited by Joseph Moxon.

ZERFASS, G. "Two Mirrors, the Story of the Invention of the Sextant", *Navigation*, **3** (1952), 131–137.

Index

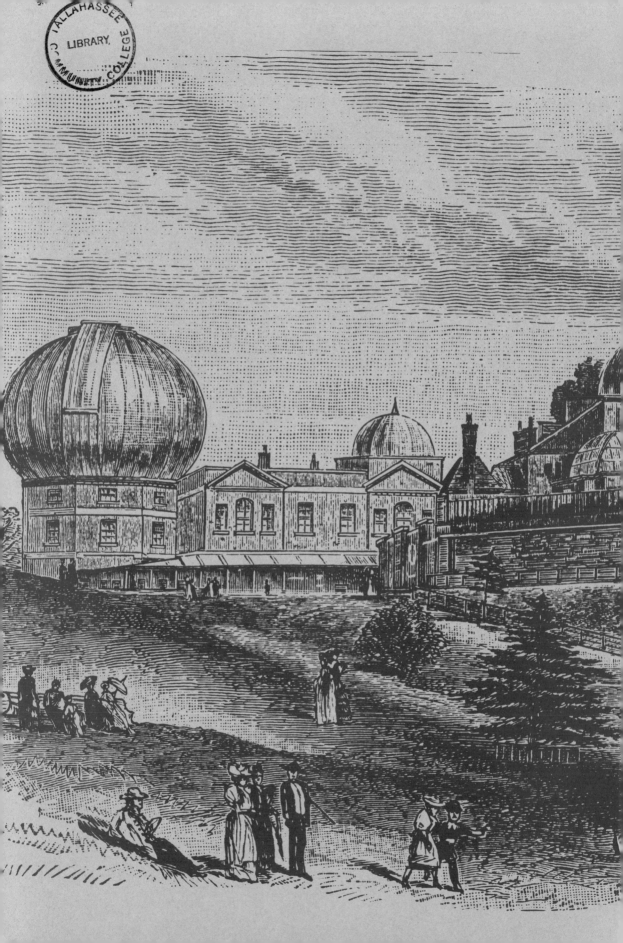